Distributed Detection and Data Fusion

Springer
*New York
Berlin
Heidelberg
Barcelona
Budapest
Hong Kong
London
Milan
Paris
Santa Clara
Singapore
Tokyo*

Signal Processing and Data Fusion

Synthetic Aperture Radar
J.P. Fitch

Multiplicative Complexity, Convolution and the DFT
M.T. Heideman

Array Signal Processing
S.U. Pillai

Maximum Likelihood Deconvolution
J.M. Mendel

Algorithms for Discrete Fourier Transform and Convolution
T. Tolimieri, M. An, and C. Lu

Algebraic Methods for Signal Processing and Communications Coding
R.E. Blahut

Electromagnetic Devices for Motion Control and Signal Processing
Y.M. Pulyer

Mathematics of Multidimensional Fourier Transform Algorithms
R. Tolimieri, M. An, and C. Lu

Lectures on Discrete Time Filtering
R.S. Bucy

Distributed Detection and Data Fusion
P.K. Varshney

Pramod K. Varshney

Distributed Detection and Data Fusion

C.S. Burrus
Consulting Editor

With 62 Illustrations

Springer

Pramod K. Varshney
Department of Electrical and
 Computer Engineering
Syracuse University
Syracuse, NY 13244-1240
USA

Consulting Editor
Signal Processing and Digital Filtering

C.S. Burrus
Professor and Chairman
Department of Electrical and
 Computer Engineering
Rice University
Houston, TX 77251-1892
USA

Library of Congress Cataloging-in-Publication Data
Varshney, Pramod K.
    Distributed detection and data fusion/Pramod K. Varshney.
        p.   cm. — (Signal processing and digital filtering)
    Includes bibliographical references and index.
    ISBN 0-387-94712-4 (hardcover: alk. paper)
    1. Signal processing.   2. Multisensor data fusion.   3. Signal
detection.   I. Title.   II. Series.
TK5102.5.V36   1996
621.382'2 – dc20   96-10600

Printed on acid-free paper.

© 1997 Springer-Verlag New York, Inc.
All rights reserved. This work may not be translated or copied in whole or in part without the written permission of the publisher (Springer-Verlag New York, Inc., 175 Fifth Avenue, New York, NY 10010, USA), except for brief excerpts in connection with reviews or scholarly analysis. Use in connection with any form of information storage and retrieval, electronic adaptation, computer software, or by similar or dissimilar methodology now known or hereafter developed is forbidden.
The use of general descriptive names, trade names, trademarks, etc., in this publication, even if the former are not especially identified, is not to be taken as a sign that such names, as understood by the Trade Marks and Merchandise Marks Act, may accordingly be used freely by anyone.

Production managed by Francine McNeill; manufacturing supervised by Jacqui Ashri.
Photocomposed copy prepared using the author's WordPerfect files.
Printed and bound by Braun-Brumfield, Inc., Ann Arbor, MI.
Printed in the United States of America.

9 8 7 6 5 4 3 2 1

ISBN 0-387-94712-4 Springer-Verlag New York Berlin Heidelberg   SPIN 10431859

*To My Parents,
Raj Kumar & Narvada Devi Varshney,
who celebrated their Golden Anniversary in 1996*

कर्मण्येवाधिकारस्ते मा फलेषु कदाचन ।
मा कर्मफलहेतुर्भूर्मा ते सङ्गोऽस्त्वकर्मणि ॥

— श्री कृष्ण
२.४७ भगवद गीता

Your right is to action alone, never to the fruits thereof. Fruits of action should not be your motive, nor should you avoid action.

— Sri Krishna
2.47 Bhagvad Gita

Preface

This book provides an introductory treatment of the fundamentals of decision–making in a distributed framework. Classical detection theory assumes that complete observations are available at a central processor for decision–making. More recently, many applications have been identified in which observations are processed in a distributed manner and decisions are made at the distributed processors, or processed data (compressed observations) are conveyed to a fusion center that makes the global decision. Conventional detection theory has been extended so that it can deal with such distributed detection problems. A unified treatment of recent advances in this new branch of statistical decision theory is presented. Distributed detection under different formulations and for a variety of detection network topologies is discussed. This material is not available in any other book and has appeared relatively recently in technical journals. The level of presentation is such that the book can be used as a graduate–level textbook. Numerous examples are presented throughout the book. It is assumed that the reader has been exposed to detection theory. The book will also serve as a useful reference for practicing engineers and researchers.

I have actively pursued research on distributed detection and data fusion over the last decade, which ultimately interested me in writing this book. Many individuals have played a key role in the completion of this book. I would like to thank Vince Vannicola for his continued interest in my research and for his helpful suggestions throughout this period. It is a pleasure to acknowledge the stimulating discussions that

I have had with my numerous doctoral students. Their contribution to this research was invaluable. Debbie Tysco started typing this manuscript. Cynthia Bromka-Skafidas gave it its final form. I am indebted to Cynthia for cheerfully going through many cycles of revisions and corrections. Chao-Tang Yu, Vajira Samarasooriya, and Mücahit Üner prepared the figures included in this book. I greatly value their help. Thanks to Maureen Marano and Vajira Samarasooriya for their help during a critical phase of this project. I am grateful to Rome Laboratory and the Air Force Office of Scientific Research (AFOSR) for sponsoring my research. Special thanks go to Vince Vannicola and Jon Sjogren in this regard. Finally, I am at a loss to find suitable words that express my deep appreciation for the significant contributions of my wife, Anju, and my sons, Lav and Kush. They have been extremely supportive throughout this project. Completion has been made possible by their constant encouragement, patience, and understanding.

Syracuse, New York   Pramod K. Varshney

Contents

Preface		ix
1	Introduction	1
	1.1 Distributed Detection Systems	1
	1.2 Outline of the Book	4
2	Elements of Detection Theory	6
	2.1 Introduction	6
	2.2 Bayesian Detection Theory	7
	2.3 Minimax Detection	14
	2.4 Neyman–Pearson Test	16
	2.5 Sequential Detection	18
	2.6 Constant False Alarm Rate (CFAR) Detection	24
	2.7 Locally Optimum Detection	32
3	Distributed Bayesian Detection: Parallel Fusion Network	36
	3.1 Introduction	36
	3.2 Distributed Detection Without Fusion	37
	3.3 Design of Fusion Rules	59
	3.4 Detection with Parallel Fusion Network	72
4	Distributed Bayesian Detection: Other Network Topologies	119
	4.1 Introduction	119
	4.2 The Serial Network	120
	4.3 Tree Networks	137

	4.4 Detection Networks with Feedback	139
	4.5 Generalized Formulation for Detection Networks	159
5	Distributed Detection with False Alarm Rate Constraints	179
	5.1 Introduction	179
	5.2 Distributed Neyman–Pearson Detection	180
	5.3 Distributed CFAR Detection	191
	5.4 Distributed Detection of Weak Signals	206
6	Distributed Sequential Detection	216
	6.1 Introduction	216
	6.2 Sequential Test Performed at the Sensors	217
	6.3 Sequential Test Performed at the Fusion Center	226
7	Information Theory and Distributed Hypothesis Testing	233
	7.1 Introduction	233
	7.2 Distributed Detection Based on Information Theoretic Criterion	234
	7.3 Multiterminal Detection with Data Compression	245
Selected Bibliography		251
Index		273

1
Introduction

1.1 Distributed Detection Systems

All of us frequently encounter decision-making problems in every day life. Based on our observations regarding a certain phenomenon, we need to select a particular course of action from a set of possible options. This problem involving a single decision maker is typically a difficult one. Decision making in large-scale systems consisting of multiple decision makers is an even more challenging problem. Group decision-making structures are found in many real world situations. Application areas include financial institutions, air-traffic control, oil exploration, medical diagnosis, military command and control, electric power networks, weather prediction, and industrial organizations. For example, a medical doctor may order multiple diagnostic tests and seek additional peer opinions before a major surgical procedure is carried out, or a military commander may use data from radar and IR sensors along with intelligence information while deciding whether or not to launch an offensive. In many applications, multiple decision makers arise naturally, e.g., managers in an industrial organization. In many other applications, additional decision makers are employed to improve system performance. For example, deployment of multiple sensors for signal detection in a military surveillance application improves system survivability, results in improved detection performance or in a shorter decision time to attain a prespecified performance level, and may provide increased coverage in terms of surveillance region and number of targets.

Many organizational structures for group decision-making systems can be envisioned. These depend upon the nature of the problem and the application domain. A variety of issues need to be examined to determine the most suitable paradigm. Some of these issues are,

1. Are the decision makers geographically or functionally distributed?

2. What is the hierarchical structure of the organization, e.g., who reports to whom?

3. Are there restrictions on communication among decision makers? What is the nature of these restrictions, e.g., how much communication bandwidth is available?

4. What are the characteristics and format of messages that are exchanged by the decision makers?

5. What is the nature and accuracy of information available to decision makers?

6. How much computational capability is available to decision makers?

7. Does the application require extensive on-line processing?

Based upon the answers to the above issues and the desired goals, an appropriate paradigm can be selected for a specific group decision-making problem. In general, decision makers coordinate their efforts to attain some system-wide goals.

In a group decision-making system, multiple sensors observe a common phenomenon. If there are no constraints on communication channel and processor bandwidths, complete observations can be brought to a central processor for data processing. In this case, sensors act as simple data collectors and do not perform any data processing. Signal processing is centralized in nature, and conventional optimal algorithms can be implemented. In many practical situations, especially when sensors are dispersed over a wide geographic area, there are limitations on the amount of communication allowed among sensors. Also, sensors are provided with processing capabilities. In this case, a certain amount of computation can be performed at the individual sensors and a compressed version of sensor data can be transmitted to a fusion center where the received information is appropriately combined to yield the global inference. The fusion center is faced with a conventional decision-making problem in that it must make a decision based on information received from the sensors. Additional issues that need to be addressed are the signal processing schemes at the individual sensors and the nature of information transmitted from the sensors to the fusion center. As an example, sensors may make hard decisions and transmit these results to the fusion center for decision combining. Other options include transmission of soft decisions (multi-level decisions instead of binary decisions) or transmission of quantized observations to the fusion center. In each case, the amount of transmission required is different. This detection network topology is widely known as a parallel fusion network. It should be pointed out at this stage that, in a parallel fusion network employing the decentralized or distributed scheme described above, the fusion center has access to less information and, consequently, system performance is not as good as a centralized scheme. But still its advantages, such as reduced communication requirements and survivability, make such distributed detection systems an attractive alternative for many applications that employ multiple sensors.

A variety of additional detection network topologies can be envisaged. Several variations of the parallel fusion network can be considered. For example, the fusion center can be allowed to make direct observations of the phenomenon. This configuration is known as a parallel fusion network with side information. In some situations, global inference is not desired. Individual sensors observe the common phenomenon, and reach local decisions while optimizing some system-wide objective function. This system topology is known as a parallel configuration without fusion. One may introduce feedback from the fusion center to the individual sensors in a parallel fusion network. In this case, observations arrive at the local detectors (sensors) sequentially. Each local detector makes a decision based on its observations and feedback information received from the fusion center. These local decisions are transmitted to the fusion center where they are combined to yield a global decision. This global decision is transmitted back to the local detectors to be used by them in their decision making. Global decision is accepted when a desired level of performance is achieved or observations cease to arrive. Another frequently used distributed detection topology is the serial or tandem topology. In this system, the first detector observes the phenomenon, makes a decision, and transmits it to the next detector. Based on the incoming decision and its own observation, the second detector computes its decision and transmits it to the next detector. This process continues until the final detector of the serial network which yields the global inference. Other important topologies, such as a tree network, will be introduced later in the book.

1.2 Outline of the Book

This book is devoted to the design and analysis of distributed detection systems. Decision rules at different detectors under a variety of detection criteria are determined. Before we deal with distributed detection problems, we present a brief review of classical detection theory in Chapter 2. Topics covered include Bayesian detection, minimax detection, Neyman–Pearson test, sequential detection, constant false alarm rate (CFAR) detectors and the locally optimum detection problem for the weak signal case.

We begin our study of distributed detection theory in Chapter 3 where we study the parallel fusion network under a Bayesian formulation. First, we consider the parallel configuration without fusion and solve binary and ternary hypothesis testing problems. Then, we obtain fusion rules when the statistics of the incoming decisions are available. Next, the design of the overall system is addressed, and decision rules at the sensors and the fusion center are derived. A number of important issues, such as computational complexity, computational algorithms, robust and nonparametric detection, are also treated.

In Chapter 4, we continue our study of distributed Bayesian detection problems for several other network topologies. The serial network and the parallel fusion network with feedback are considered in detail. Towards the end, a unified methodology to represent different network topologies is presented, and decision rules at all the detectors are obtained.

Chapter 5 considers distributed detection for scenarios where constraints are specified on the probability of false alarm, i.e., where it is to remain less than an acceptable value. First, the Neyman–Pearson formulation of the problem is considered. Decision rules are obtained, so that the overall probability of false alarm remains under a given value and probability of detection is maximized. Distributed CFAR detection based on adaptive thresholding for nonstationary environments is considered. Finally, distributed detection of weak signals is addressed. Attention is limited to the parallel fusion network in this chapter.

Sequential methods are applied to distributed detection problems in Chapter 6. A Bayesian formulation of two distributed sequential detection problems for parallel fusion network topologies is considered. First, a two-sensor parallel network without fusion, in which sequential tests are implemented at the sensors, is studied. Then, a parallel fusion network, in which the sequential test is implemented at the fusion center, is treated.

In Chapter 7, some results on information theory applications to the distributed detection problem are briefly presented. First, design of a parallel fusion network based on an information theoretic cost function is discussed. Then, some asymptotic results on the performance of distributed detection systems are briefly described.

2
Elements of Detection Theory

2.1 Introduction

There are many practical situations in which one is faced with a decision-making problem, i.e., the problem of choosing a course of action from several possibilities. For example, in a radar detection context, a decision is to be made regarding the presence or absence of a target based on the radar return. In a digital communication system, one of several possible waveforms is transmitted over a channel. Based on the received noisy observation, we need to determine the symbol that was transmitted. In a biomedical application, based on a smear of human tissue, one needs to determine if it is cancerous. In a pattern recognition problem, one needs to determine the type of aircraft being observed based on some aircraft features. In all of the above applications, the common underlying problem is to make a decision among several possible choices. This is carried out based on available noisy observations. The branch of statistics dealing with these types of problems is known as statistical decision theory or hypothesis testing. In the context of radar and communication theory, it is known as detection theory.

In decision-making problems, we may be faced with a binary problem, i.e., a problem with a yes or no type of answer, or we may have several choices. These different possibilities are referred to as hypotheses. In a binary hypothesis testing problem, the two hypotheses are denoted by H_0 and H_1. Hypothesis H_0 usually represents the transmission of symbol zero or the absence of a target. Hypothesis H_1

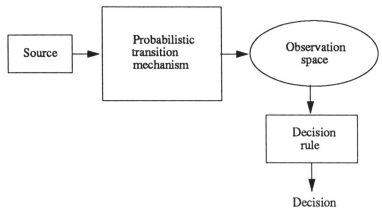

Figure 2.1. Components of a hypothesis testing problem.

corresponds to the transmission of symbol one or the presence of a target. In the more general situation with multiple hypotheses, the hypotheses are denoted by $H_0, H_1, ..., H_{M-1}$. In this book, we will concentrate mostly on binary hypothesis testing problems. The more general case of M-ary hypothesis testing can be treated similarly.

The basic components of a hypothesis testing problem are shown in Figure 2.1. The source generates one of the hypotheses as an output. The source output is not directly observable, otherwise there would be no decision problem. The probabilistic transition mechanism knows which hypothesis is true and generates a point in the observation space according to some probability law. We have access to the observation from which we have to make a decision on the hypothesis that is present. To accomplish this, we derive a decision rule that partitions the observation space into decision regions corresponding to the different hypotheses. The hypothesis, corresponding to the decision region where the observation falls, is declared true. The decision rule depends upon the optimization criterion in use. Derivation of decision rules based on different criteria will be discussed in the rest of this chapter.

2.2 Bayesian Detection Theory

Let us consider the simple binary hypothesis testing problem with the two hypotheses denoted by H_0 and H_1. Let the observation be denoted as y so that the conditional densities under the two hypotheses are $p(y|H_i)$, $i = 0, 1$. These are assumed known, and the points in the observation space are generated in accordance with these conditional densities. The a priori probabilities of the two hypotheses are denoted by P_0 and P_1 respectively. In the binary hypothesis testing problem, four possible courses of action can occur. Two of these correspond to correct decisions and two to errors. We assign costs to each of these possibilities. Specifically, C_{ij}, $i = 0, 1, j = 0, 1$, represents the cost of declaring H_i true when H_j is present. In the Bayesian formulation, the decision rule that minimizes the average cost is determined. The average cost or Bayes risk function, $\Re$, is given by

$$\Re = \sum_{i=0}^{1} \sum_{j=0}^{1} C_{ij} P_j P(\text{Decide } H_i | H_j \text{ is present})$$

$$= \sum_{i=0}^{1} \sum_{j=0}^{1} C_{ij} P_j \int_{Z_i} p(y|H_j) dy, \qquad (2.2.1)$$

where Z_i is the decision region corresponding to hypothesis H_i, i.e., hypothesis H_i is declared true for any observation falling in the region Z_i. Let Z be the entire observation space so that $Z = Z_0 \cup Z_1$ and $Z_0 \cap Z_1 = \phi$, the null set. We may expand (2.2.1) as

$$\Re = P_0 C_{00} \int_{Z_0} p(y|H_0) \, dy + P_0 C_{10} \int_{Z-Z_0} p(y|H_0) \, dy$$

$$+ P_1 C_{01} \int_{Z_0} p(y|H_1) \, dy + P_1 C_{11} \int_{Z-Z_0} p(y|H_1) \, dy. \qquad (2.2.2)$$

Noting that

$$\int_Z p(y|H_j) \, dy = 1, \quad j = 0, 1$$

and collecting terms, (2.2.2) reduces to

$$\Re = P_0 C_{10} + P_1 C_{11} + \int_{Z_0} \{[P_1(C_{01} - C_{11}) p(y|H_1)] - [P_0(C_{10} - C_{00}) p(y|H_0)]\} dy. \qquad (2.2.3)$$

The first two terms on the right hand side are fixed. The risk $\Re$ can be minimized by assigning those points of Z to Z_0 that make the integrand of (2.2.3) negative. Assuming $C_{10} > C_{00}$ and $C_{01} > C_{11}$, the minimization results in the likelihood ratio test (LRT)

$$\frac{p(y|H_1)}{p(y|H_0)} \begin{array}{c} H_1 \\ > \\ < \\ H_0 \end{array} \frac{P_0(C_{10} - C_{00})}{P_1(C_{01} - C_{11})}. \qquad (2.2.4)$$

The quantity on the left hand side is known as the likelihood ratio denoted by $\Lambda(y)$ and the quantity on the right hand side is the threshold η, i.e.,

$$\Lambda(y) = \frac{p(y|H_1)}{p(y|H_0)},$$

$$\eta = \frac{P_0(C_{10} - C_{00})}{P_1(C_{01} - C_{11})}.$$

The LRT can then be expressed as

$$\Lambda(y) \begin{array}{c} H_1 \\ > \\ < \\ H_0 \end{array} \eta. \qquad (2.2.5)$$

Because the natural logarithm is a monotonically increasing function and the two sides of the LRT are positive, an equivalent test is

$$\log \Lambda(y) \begin{array}{c} H_1 \\ > \\ < \\ H_0 \end{array} \log \eta. \qquad (2.2.6)$$

This form of the test in terms of the log likelihood ratio is often easier to implement.

A special case of the Bayes test occurs when we set $C_{00} = C_{11} = 0$ and $C_{10} = C_{01} = 1$, i.e., the cost of a correct decision is set to zero and the cost of an error is set to unity. In this case,

$$\Re = P_0 \int_{Z_1} p(y|H_0) \, dy + P_1 \int_{Z_0} p(y|H_1) \, dy, \quad (2.2.7)$$

which is just the average probability of error. In this case, the Bayes test simply minimizes the average probability of error. The threshold η, in this case, is given by P_0/P_1. When the two hypotheses are equally likely, $\eta = 1$, and the log likelihood ratio test utilizes a threshold equal to zero. These assumptions are usually valid in digital communication systems, and the resulting receivers are called minimum probability of error receivers.

As indicated earlier, two types of errors can occur in decision making. When H_0 is true and H_1 is declared true, an error of the first kind is said to occur. In radar terminology, it is known as a false alarm, and the associated conditional probability of error is known as the probability of false alarm, P_F. An error of the second kind, a miss, is said to occur if H_0 is declared true when H_1 is present. The associated conditional probability of error is known as the probability of miss, P_M. Then,

$$P_F = P(\text{Decide } H_1 | H_0 \text{ present}) = \int_{Z_1} p(y|H_0) \, dy, \quad (2.2.8)$$

and

$$P_M = P(\text{Decide } H_0 | H_1 \text{ present}) = \int_{Z_0} p(y|H_1) \, dy. \quad (2.2.9)$$

We also define the probability of detection P_D as the probability of declaring H_1 true when H_1 is present, i.e.,

$$P_D = 1 - P_M = \int_{Z_1} p(y|H_1) \, dy. \quad (2.2.10)$$

Based on the previous definitions, the average probability of error in decision making is given by

$$P(\text{error}) = P_0 P_F + P_1 P_M. \qquad (2.2.11)$$

In terms of P_F and P_M, the Bayes risk $\Re$ can be expressed as

$$\Re = P_0 C_{10} + P_1 C_{11} + P_1 (C_{01} - C_{11}) P_M - P_0 (C_{10} - C_{00})(1 - P_F). \qquad (2.2.12)$$

Because $P_0 = 1 - P_1$, $\Re$ in (2.2.12) becomes

$$\begin{aligned}\Re &= C_{00}(1 - P_F) + C_{10} P_F \\ &\quad + P_1[(C_{11} - C_{00}) + (C_{01} - C_{11}) P_M - (C_{10} - C_{00}) P_F].\end{aligned} \qquad (2.2.13)$$

The minimum value of $\Re$ can be determined by substituting the values of P_F and P_M corresponding to the optimum decision regions. In the following, we present an alternate approach to determine the minimum value of $\Re$.

Consider the expression for $\Re$ given in (2.2.1). From Bayes rule,

$$P_j p(y|H_j) = P(H_j|y) p(y), \qquad (2.2.14)$$

where the unconditional density function of y is given by

$$p(y) = P_0\, p(y|H_0) + P_1\, p(y|H_1). \qquad (2.2.15)$$

Using (2.2.14), we express (2.2.1) as

$$\Re = \sum_{i=0}^{1} \sum_{j=0}^{1} C_{ij} \int_{Z_i} P(H_j|y) p(y)\, dy. \qquad (2.2.16)$$

Interchanging the order of integration and inner summation,

2. Elements of Detection Theory

$$\mathcal{R} = \sum_{i=0}^{1} \int_{Z_i} \sum_{j=0}^{1} C_{ij} P(H_j|y) p(y) dy$$

$$= \sum_{i=0}^{1} \int_{Z_i} \beta_i(y) p(y) dy, \qquad (2.2.17)$$

where

$$\beta_i(y) = \sum_{j=0}^{1} C_{ij} P(H_j|y) \qquad (2.2.18)$$

are the conditional costs assigned to each point y in the observation space. The optimum receiver that minimizes the Bayes risk $\mathcal{R}$ uses the decision rule

$$\beta_0(y) \underset{H_0}{\overset{H_1}{\gtrless}} \beta_1(y). \qquad (2.2.19)$$

Let $r(y)$ denote the conditional cost of the optimum receiver. Then,

$$r(y) = \min\,[\beta_0(y), \beta_1(y)]. \qquad (2.2.20)$$

Using the mathematical identity,

$$\min\,(a,b) = \tfrac{1}{2}(a+b) - \tfrac{1}{2}|a-b|, \qquad (2.2.21)$$

we may express $r(y)$ as

$$r(y) = \tfrac{1}{2}[\beta_0(y) + \beta_1(y)] - \tfrac{1}{2}|\beta_0(y) - \beta_1(y)|. \qquad (2.2.22)$$

Using the definitions of $\beta_0(y)$ and $\beta_1(y)$ and the Bayes rule

$$P(H_j|y) = \frac{P_j\, p(y|H_j)}{p(y)},$$

we can express $r(y)$, from (2.2.22), as

$$r(y) = \frac{1}{2p(y)}\Big[P_0(C_{00}+C_{10})p(y|H_0) + P_1(C_{01}+C_{11})p(y|H_1)$$

$$- |P_1(C_{01}-C_{11})p(y|H_1) - P_0(C_{10}-C_{00})p(y|H_0)|\Big]. \quad (2.2.23)$$

The minimum value of the Bayes risk $\Re_{min}$ is obtained using (2.2.17) and (2.2.20), i.e.,

$$\Re_{min} = \int_Z r(y)p(y)\,dy, \quad (2.2.24)$$

where Z is the entire observation space. Substituting (2.2.23) into (2.2.24) and simplifying,

$$\Re_{min} = C_0 - \frac{1}{2}\int_Z |(C_{01}-C_{11})P_1 p(y|H_1) - (C_{10}-C_{00})P_0 p(y|H_0)|\,dy, \quad (2.2.25)$$

where

$$C_0 = \frac{1}{2}(C_{00}+C_{10})P_0 + \frac{1}{2}(C_{01}+C_{11})P_1.$$

In the special case, $C_{00} = C_{11} = 0$, $C_{01} = C_{10} = 1$, $\Re_{min}$ reduces to

$$\Re_{min} = \frac{1}{2} - \frac{1}{2}\int_Z |P_1 p(y|H_1) - P_0 p(y|H_0)|\,dy. \quad (2.2.26)$$

This provides the expression for the minimum achievable probability of error by the optimum Bayesian detection system. It is known as the Kolmogorov variational distance [Kai67]. If the observation y is discrete, the corresponding expression for $\Re_{min}$ is

$$\Re_{min} = C_0 - \frac{1}{2}\sum_{\text{all } y_k} |(C_{01} - C_{11})P_1 p(y_k|H_1) - (C_{10} - C_{00})P_0 p(y_k|H_0)|,$$

(2.2.27)

where y_k are the discrete values taken by the random variable y. These expressions will be found useful in the sequel.

The performance of LRTs can be conveniently described in terms of a graph known as the receiver operating characteristic (ROC). This is a plot of the probability of detection P_D against the probability of false alarm P_F. As shown in [Van68, Poo88, and Hel95a], the ROCs of all continuous LRTs are concave downward and lie above the $P_D = P_F$ line.

2.3 Minimax Detection

The Bayes criterion for the design of decision rules requires the knowledge of a priori probabilities which may not be readily available. In such situations, the minimax criterion is a feasible alternative. Under this criterion, one uses the Bayes decision rule corresponding to the least favorable prior probability assignment.

Consider the Bayes risk $\Re$ given in (2.2.13). It is a function of the prior probability P_1 and the decision regions through P_M and P_F. For a fixed cost assignment, the optimum Bayes threshold and the risk $\Re_{opt}$ vary with P_1. The risk $\Re_{opt}$ for the optimal Bayes test is a continuous concave downward function of P_1 for $P_1 \in [0, 1]$. A typical $\Re_{opt}$ versus P_1 curve is shown in Figure 2.2. It may have an interior maximum, or a maximum may occur at one of the end points. Let us now assume that a fixed value of P_1, say P_1', is chosen and a Bayes test is designed corresponding to P_1'. Let us fix the threshold at the resulting value (the one obtained by fixing P_1 at P_1'). By fixing the threshold, we also fix P_F and P_M. In this situation, the Bayes risk $\Re'$ is a linear function of P_1 as shown in Figure 2.2, i.e., if the value of P_1 deviates from its design value of P_1', the resulting $\Re'$ can be obtained from the linear function (2.2.13). This straight line is a tangent to the optimum Bayes curve because at $P_1 = P_1'$ both risks $\Re_{opt}$ and $\Re'$ attain the same value, and, at other values of P_1, $\Re' > \Re_{opt}$. Thus, the actual values of the risk $\Re'$ could be much larger than the corresponding $\Re_{opt}$. If we were to design

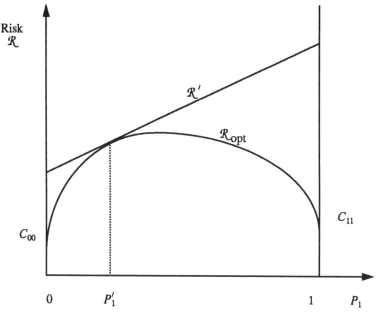

Figure 2.2. A typical $\mathcal{R}_{opt}$ versus P_1 curve.

the optimum Bayes test using $P_1 = P_1^*$ where P_1^* is the value of P_1 at which $\mathcal{R}_{opt}$ has an interior maximum, the tangent will be horizontal. In this case, no matter what the value of P_1 is, the resulting risk will be the same. Even though $\mathcal{R}_{opt}$ is maximum at the point $P_1 = P_1^*$, we are guaranteed that the risk will not exceed this value for any other value of P_1. This criterion, where the system is designed to minimize the maximum risk, is known as the minimax criterion, and the resulting decision rule is known as the minimax test. For a minimax test, the line given by (2.2.13) must be horizontal, i.e., its slope must be zero and

$$(C_{11} - C_{00}) + (C_{01} - C_{11})P_M - (C_{10} - C_{00})P_F = 0. \qquad (2.3.1)$$

This equation is known as the minimax equation and determines the test if the maximum of the risk is interior to the interval (0, 1). If the maximum occurs at the boundary, i.e., at $P_1 = 0$ or $P_1 = 1$, this value of P_1 at the boundary is used for the design of the decision rule. In the special case, $C_{00} = C_{11} = 0$, the minimax equation reduces to

$$C_{01}P_M = C_{10}P_F, \qquad (2.3.2)$$

and the risk in this case is independent of the prior probability P_1. Note that, in the above development, $\Re_{opt}$ has been assumed to be differentiable everywhere.

2.4 Neyman–Pearson Test

In many practical situations, not only are the prior probabilities unknown, but the cost assignments are also difficult to make. For example, in radar detection, the cost of missing a target cannot be determined easily. In this case, we employ the Neyman–Pearson criterion and work with the probability of false alarm P_F and the probability of detection P_D while designing the decision rule. In general, it is desirable to make P_F as small as possible and P_D as large as possible. But, these are conflicting objectives. In the Neyman–Pearson formulation, we constrain P_F to an acceptable value and maximize P_D or equivalently minimize P_M. The resulting test is known as the Neyman–Pearson test.

Let α denote the acceptable value of P_F (level of the test). The objective is to design a nonrandomized test that maximizes P_D (or equivalently minimizes P_M) under the constraint on P_F. The fundamental Neyman–Pearson lemma specifies the structure of the most powerful (maximum P_D) test [Leh86].

(i) *Existence*: For testing H_0 against H_1, a test Φ and a constant k exist, so that

$$E_{H_0}\Phi(y) = \alpha \qquad (2.4.1)$$

and

$$\Phi(y) = \begin{cases} 1, & \text{when } p(y|H_1) > kp(y|H_0), \\ 0, & \text{when } p(y|H_1) < kp(y|H_0). \end{cases} \qquad (2.4.2)$$

(ii) *Sufficient Condition*: A test that satisfies (2.4.1) and (2.4.2) for some k is most powerful for testing H_0 against H_1 at level α.

2.4 Neyman–Pearson Test

(iii) *Necessary Condition*: If Φ is most powerful at level α for testing H_0 against H_1, then for some k it satisfies (2.4.2) almost everywhere. It also satisfies (2.4.1) unless a test of level less than α and with power one exists.

A popular approach to obtain the solution is to employ the method of Lagrange multipliers. We construct the function F

$$F = P_M + \lambda [P_F - \alpha], \qquad (2.4.3)$$

where $\lambda \geq 0$ is the Lagrange multiplier. Substituting for P_M and P_F,

$$F = \int_{Z_0} p(y|H_1) dy + \lambda [\int_{Z_1} p(y|H_0) dy - \alpha]$$

$$= \lambda(1-\alpha) + \int_{Z_0} [p(y|H_1) - \lambda p(y|H_0)] dy. \qquad (2.4.4)$$

The first term is fixed, and the second term is minimized if we employ the following likelihood ratio test (LRT):

$$\Lambda(y) = \frac{p(y|H_1)}{p(y|H_0)} \underset{H_0}{\overset{H_1}{\gtrless}} \lambda. \qquad (2.4.5)$$

The threshold of the test is the Lagrange multiplier λ chosen to satisfy the constraint

$$P_F = \int_{Z_1} p(y|H_0) dy = \int_\lambda^\infty p(\Lambda|H_0) d\Lambda = \alpha. \qquad (2.4.6)$$

The solution of (2.4.6) yields the desired threshold.

A more general formulation of the Neyman–Pearson test allows randomization and has the structure

$$P(\text{Decide } H_1) = \begin{cases} 0, & \text{if } \Lambda(y) < \lambda, \\ \gamma, & \text{if } \Lambda(y) = \lambda, \\ 1, & \text{if } \Lambda(y) > \lambda. \end{cases} \quad (2.4.7)$$

In this case, threshold λ and randomization probability γ that maximize P_D and attain level $P_F = \alpha$ are evaluated.

2.5 Sequential Detection

In the previous sections, we have considered detection problems based on scalar observations. The same methodology is employed for a multiple but fixed number of observation samples. In many practical situations, however, observations are collected sequentially, and more information becomes available as time progresses. In such cases, we may wish to process the observations sequentially and make a final decision as soon as we are satisfied with the decision quality or detection performance. In a radar detection problem, we may transmit pulses and observe returns until satisfied with the amount of collected information and resulting detection performance. The aim is to take additional observations only if they are necessary. In the sequential decision process, after each observation, the receiver computes the likelihood ratio and compares it with two thresholds. Either it decides on one of the two hypotheses or it decides to take another observation. The main advantage of sequential hypothesis testing is that it requires, on an average, fewer observations to achieve the same probability of error performance as a fixed-sample-size test. This advantage is attained at the expense of additional computation. A major drawback of this approach is that the time to reach a decision is random and buffering is needed for practical implementation.

Here we concentrate on Wald's sequential test or the sequential probability ratio test (SPRT). This is a Neyman–Pearson type of test where the thresholds are determined by the specified probabilities of error. In the SPRT, at each stage of the test, the likelihood ratio is computed and compared with two thresholds η_0 and η_1 which are determined by the specified values of P_F and P_M. If the likelihood ratio is greater than or equal to η_1, we decide that H_1 is present. If the likelihood ratio is less than or equal to η_0, we decide that H_0 is present. Otherwise, the decision is to take another observation because sufficient amount of information is not available to make the decision that satisfies the specified values of P_F and P_M.

Let $y_1, y_2, ..., y_k, ...$ represent the observations where y_k denotes the kth observation received sequentially. Let y_K represent the vector consisting of observations received up to and including time K,

$$y_K^T = [y_1, y_2, ..., y_K]. \qquad (2.5.1)$$

The likelihood ratio at time K is defined as

$$\Lambda(y_K) = \frac{p(y_K|H_1)}{p(y_K|H_0)}. \qquad (2.5.2)$$

Computation of $\Lambda(y_K)$ at each stage requires the knowledge of the joint conditional density functions of the observations $y_1, y_2, ..., y_K$. If the observations are assumed to be independent, this computation simplifies considerably and can be done recursively. In this case,

$$\Lambda(y_K) = \frac{p(y_K|H_1)}{p(y_K|H_0)} = \prod_{k=1}^{K} \frac{p(y_k|H_1)}{p(y_k|H_0)}$$

$$= \frac{p(y_K|H_1)}{p(y_K|H_0)} \prod_{k=1}^{K-1} \frac{p(y_k|H_1)}{p(y_k|H_0)}$$

$$= \Lambda(y_K) \Lambda(y_{K-1}). \qquad (2.5.3)$$

The likelihood ratio for the first observation is defined as

$$\Lambda(y_1) = \Lambda(y_1) = \frac{p(y_1|H_1)}{p(y_1|H_0)}. \qquad (2.5.4)$$

The next step is to determine the thresholds η_0 and η_1 in terms of the specified values of P_F and P_M. Let us assume that the two error probabilities are specified as

$$P_F = \alpha, \quad (2.5.5)$$

and

$$P_M = \beta. \quad (2.5.6)$$

The sequential test is implemented as

$$\Lambda(y_K) \geq \eta_1 \quad \text{decide } H_1, \quad (2.5.7a)$$
$$\Lambda(y_K) \leq \eta_0 \quad \text{decide } H_0. \quad (2.5.7b)$$

Otherwise, the test is continued by taking another observation. Recall that P_F and P_D are given by

$$P_F = \int_{Z_1} p(y_K | H_0) dy_K \quad (2.5.8)$$

and

$$P_D = \int_{Z_1} p(y_K | H_1) dy_K$$
$$= \int_{Z_1} \Lambda(y_K) p(y_K | H_0) dy_K, \quad (2.5.9)$$

where Z_1 is the decision region corresponding to hypothesis H_1. To decide H_1, we must have $\Lambda(y_K) \geq \eta_1$. Therefore, we have the following inequality in terms of P_D

$$P_D \geq \eta_1 \int_{Z_1} p(y_K | H_0) dy_K. \quad (2.5.10)$$

Because $P_D = 1 - P_M = 1 - \beta$, using (2.5.8) we may express (2.5.10) as

2.5 Sequential Detection

$$1 - \beta \geq \eta_1 \alpha, \quad (2.5.11)$$

and the threshold η_1 satisfies

$$\eta_1 \leq \frac{1-\beta}{\alpha}. \quad (2.5.12)$$

Now we make an approximation. We assume that when H_1 is decided, $\Lambda(y_K) = \eta_1$, i.e., the likelihood ratio is exactly equal to the threshold. It is a reasonable assumption because, at the decision stage, the likelihood ratio is likely to exceed the threshold only by a small amount. With this approximation,

$$\eta_1 = \frac{1-\beta}{\alpha}. \quad (2.5.13)$$

Similarly, we can obtain

$$\eta_0 = \frac{\beta}{1-\alpha}. \quad (2.5.14)$$

Once again, the log likelihood ratio is often attractive from a computational standpoint. Let $L(y_K)$ denote the log likelihood ratio, i.e.,

$$L(y_K) = \log \Lambda(y_K). \quad (2.5.15)$$

Then, in terms of the log likelihood ratios, (2.5.3) can be expressed as

$$L(y_K) = L(y_K) + L(y_K - 1), \quad (2.5.16)$$

and the corresponding thresholds are

$$\log \eta_1 = \log \left(\frac{1-\beta}{\alpha} \right), \quad (2.5.17)$$

$$\log \eta_0 = \log\left(\frac{\beta}{1-\alpha}\right). \tag{2.5.18}$$

It was mentioned previously that the number of observations required for terminating the test is random. We compute the average number of observations required for terminating the test under each hypothesis. Let us assume that the test terminates at the Kth stage. This implies that, at the Kth stage, the likelihood ratio is equal to $\log \eta_0$ or $\log \eta_1$. If hypothesis H_0 is present, and $L(y_K) \geq \log \eta_1$, a false alarm occurs with probability α. In this case, the probability that $L(y_K) \leq \log \eta_0$ is $1 - \alpha$. Therefore, we may express the expected value of $L(y_K)$ under hypothesis H_0 as

$$E[L(y_K)|H_0] = \alpha \log \eta_1 + (1-\alpha) \log \eta_0. \tag{2.5.19}$$

Similarly,

$$E[L(y_K)|H_1] = (1-\beta) \log \eta_1 + \beta \log \eta_0. \tag{2.5.20}$$

Let us define an indicator variable I_k as

$$I_k = \begin{cases} 1, & \text{if no decision made up to the } (k-1)\text{st stage}, \\ 0, & \text{if decision made at an earlier stage}. \end{cases} \tag{2.5.21}$$

Because the test terminates at the Kth stage, we may express the log likelihood ratio in terms of the indicator variable as

$$L(y_K) = \sum_{k=1}^{K} L(y_k) = \sum_{k=1}^{\infty} I_k L(y_k). \tag{2.5.22}$$

Note that I_k depends only on $y_1, \ldots, y_{k-1}$ and not on y_k. Therefore, I_k is independent of $L(y_k)$. Taking the expectation on both sides of (2.5.22), under the two hypotheses,

$$E[L(y_K)|H_i] = E\left[\sum_{k=1}^{\infty} I_k L(y_k)|H_i\right]$$

$$= \sum_{k=1}^{\infty} E[I_k|H_i]E[L(y_k)|H_i], \quad i = 0, 1. \quad (2.5.23)$$

Let us assume that the observations y_k, $k=1,..., K$, are identically distributed, i.e.,

$$E[L(y_k)|H_i] = E[L(y)|H_i], \quad k = 1, ..., K; \quad i = 0, 1. \quad (2.5.24)$$

Then,

$$E[L(y_K)|H_i] = E[L(y)|H_i]\sum_{k=1}^{\infty} E[I_k|H_i], \quad i = 0, 1. \quad (2.5.25)$$

From the definition of the indicator variable I_k,

$$E[I_k|H_i] = 0 \times P(K < k|H_i) + 1 \times P(K \geq k|H_i)$$

$$= P(K \geq k|H_i), \quad i = 0, 1. \quad (2.5.26)$$

Therefore,

$$\sum_{k=1}^{\infty} E[I_k|H_i] = \sum_{k=1}^{\infty} P(K \geq k|H_i) = \sum_{k=1}^{\infty} kP(K = k|H_i)$$

$$= E[K|H_i]. \quad (2.5.27)$$

Substituting (2.5.27) in (2.5.25),

$$E[K|H_i] = \frac{E[L(y_K)|H_i]}{E[L(y)|H_i]}, \quad i = 0, 1. \quad (2.5.28)$$

Substituting from (2.5.19) and (2.5.20),

$$E[K|H_1] = \frac{(1-\beta)\log\eta_1 + \beta\log\eta_0}{E[L(y)|H_1]} \qquad (2.5.29)$$

and

$$E[K|H_0] = \frac{\alpha\log\eta_1 + (1-\alpha)\log\eta_0}{E[L(y)|H_0]}. \qquad (2.5.30)$$

It has been shown by Wald [Wal47] that the sequential test terminates with probability one. Moreover, it has been proven that, for specified values of P_F and P_M, the SPRT minimizes the average number of observations $E[K|H_0]$ and $E[K|H_1]$. In our development, we assumed that the observations y_k were independent and identically distributed. Without the independence assumption, the likelihood ratio cannot be computed in a recursive manner. Another assumption was that the probabilities P_F and P_M were stage invariant. If this was not the case, the thresholds would not be stage invariant. Even though it has been shown that the SPRT terminates with probability one, the number of observations required prior to termination may be large. In a practical implementation, the SPRT is truncated after a certain prespecified number of observations have been processed. At the truncation stage, the detector is forced to make a decision in favor of H_0 or H_1. The resulting test is known as a truncated SPRT and is suboptimum.

2.6 Constant False Alarm Rate (CFAR) Detection

Next, we consider signal detection in practical radar systems, i.e., detection of targets immersed in a noise and clutter background. Clutter refers to undesired radar signal returns from scatterers that are not of interest to the user. Surveillance volume is divided into small regions called range cells (in terms of distance from the radar), and the problem is to determine whether or not there is a target in a given range cell (test cell). Here, we confine our attention to the case where background noise and clutter are Gaussian. In radar signal detection, the usual objective is to maintain a constant probability of false alarm. When the noise and clutter background is stationary and its probability distribution along with its parameters is completely known, the Neyman–Pearson detector can be employed as it maintains a constant probability of false alarm and at the same time maximizes the probability of detection. In a nonstationary background, however, a fixed-threshold Neyman–Pearson detector cannot be used because, as the background conditions vary, the resulting value of P_F may be too high or the value of P_D may be too low. Therefore, adaptive threshold techniques are employed. In the Gaussian case under consideration, the adaptive threshold is based on the estimate of the mean power level of the background obtained from the range cells surrounding the test cell. A block diagram of the CFAR processor is shown in Figure 2.3. The square-law detected range samples are sent serially into a shift register of length $(N + 1)$. The leading $N/2$ samples and the lagging $N/2$ samples surrounding the test cell form the reference window. The estimate of the background noise power, Z, is obtained from the reference window samples. Two algorithms for the computation of Z will be described later in this section. The statistic Z is multiplied by a scale factor or threshold multiplier T so as to maintain P_F at a desired constant level. The product TZ is the resulting adaptive threshold. The test cell sample, Y, from the center tap is then compared with this adaptive threshold to make the decision.

Here, we consider a special case of target and noise models in more detail. We assume that the target in the test cell (primary target) is a slowly fluctuating target of Swerling type I and the background noise is Gaussian. In this case, the output of the square-law detector is exponentially distributed with probability density function (pdf)

$$p(x) = (1/2\lambda)\exp(-x/2\lambda), \quad x \geq 0. \qquad (2.6.1)$$

Under hypothesis H_0, corresponding to no target in the test cell and a homogeneous background, λ is the total background clutter-plus-thermal noise power which is denoted by σ. Under hypothesis H_1, corresponding to the presence of the target in the test cell, λ is equal to $\sigma(1 + S)$, where S is the average signal-to-noise ratio, SNR, of the primary target. For the reference window cells, λ is assumed to be equal to σ. Thus,

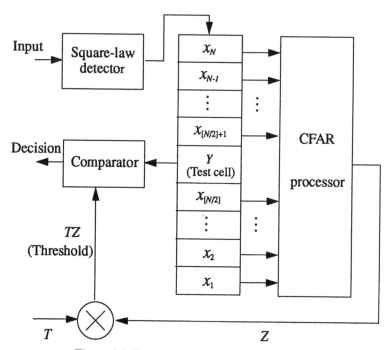

Figure 2.3. Block diagram of a CFAR processor.

$$\lambda = \begin{cases} \sigma, & \text{under } H_0 \\ \sigma(1+S), & \text{under } H_1. \end{cases} \quad (2.6.2)$$

We also assume that the observations corresponding to $(N + 1)$ cells, including the test cell, are statistically independent.

The optimum Neyman–Pearson detector assumes that the homogeneous background noise distribution and, hence, the total noise power σ is known a priori and sets an optimum fixed threshold. With this assumption, P_F is given by

$$P_F = P(Y > T_O | H_0) = \exp(-T_O/2\sigma), \quad (2.6.3)$$

where T_O denotes the optimum fixed threshold. The probability of detection corresponding to T_O, P_D^{opt}, is given by

$$P_D^{opt} = P(Y > T_0 | H_1) = \exp[-T_0/2\sigma(1+S)]. \qquad (2.6.4)$$

From (2.6.3) and (2.6.4), we can obtain the relationship

$$P_D^{opt} = (P_F)^{1/(1+S)}. \qquad (2.6.5)$$

The optimum Neyman–Pearson detector has the best performance that can be achieved for the given target and noise models. The CFAR processor with an adaptive threshold has an inherent loss of detection probability when compared with the optimum processor. This is due to the fact that the threshold in the CFAR processor is set by estimating the total noise power from only a finite number of observed noise samples whereas, in the optimum detector, the threshold is set by assuming that the total noise power is known a priori which is equivalent to saying that estimation is done over an infinite number of noise samples.

In a CFAR processor, the false alarm probability is given by

$$P_F = \int_0^\infty P(Y > TZ | Z, H_0) p_Z(z) \, dz, \qquad (2.6.6)$$

which can also be written as

$$P_F = E_Z[P(Y > TZ | H_0)], \qquad (2.6.7)$$

where Z is the estimated homogeneous background noise power level, $p_Z(z)$ is the pdf of Z and $E_Z(\cdot)$ is the expectation over Z. By inserting the pdf of Y into (2.6.7),

$$\begin{aligned} P_F &= E_Z\left[\int_{TZ}^\infty p(y|H_0) dy\right] \\ &= E_Z\left[\int_{TZ}^\infty (1/2\sigma)\exp(-y/2\sigma) dy\right] \\ &= E_Z\left[\exp(-TZ/2\sigma)\right] \\ &= M_Z(T/2\sigma), \qquad (2.6.8) \end{aligned}$$

where $M_Z(\cdot)$ denotes the moment generating function (mgf) of the random variable Z. For the computation of the moment generating function of the random variable Z, the pdf of Z is needed. Once the pdf of Z is available, the value of scale factor T for a given false alarm probability can be computed iteratively.

Similarly, the probability of detection is given by

$$P_D = E_Z[P(Y > TZ | H_1)], \qquad (2.6.9)$$

which can be rewritten as

$$\begin{aligned} P_D &= E_Z\left[\int_{TZ}^{\infty} p(y|H_1) dy\right] \\ &= E_Z\left[\int_{TZ}^{\infty} [1/2\sigma(1+S)] \exp[-y/2\sigma(1+S)] dy\right] \\ &= E_Z\left[\exp[-TZ/2\sigma(1+S)]\right] \\ &= M_Z[T/2\sigma(1+S)]. \qquad (2.6.10) \end{aligned}$$

It must be noted that, in the design of a CFAR processor, it is assumed that the background is homogeneous and the scale factor T is computed for the given noise model, the probability of false alarm level P_F, reference window size N, the CFAR algorithm, and (if applicable) its parameters. Next, we discuss two popular approaches for the computation of Z.

Cell Averaging CFAR Processor

In the cell averaging CFAR (CA-CFAR) processor, the statistic Z is set equal to the maximum likelihood estimate of the noise power computed from the reference window samples X_i, $i=1, \ldots, N$, i.e.,

$$Z = \sum_{i=1}^{N} X_i. \qquad (2.6.11)$$

2.6 CFAR Detection

When the background is homogeneous and the reference window contains independent and identically distributed observations (exponential distribution), the CA-CFAR processor attains the maximum value of detection probability P_D. As the length of the reference window increases, the P_D of this system approaches that of the classical Neyman–Pearson processor in which the background noise power is assumed to be known a priori.

The values of P_F and P_D of the CA-CFAR processor can be computed by noting that the exponential pdf is a special case of the Gamma pdf when we set $\alpha = 1$ in the following pdf expression:

$$p(v) = \beta^{-\alpha} v^{\alpha-1} \exp(-v/\beta)/\Gamma(\alpha), \quad v \geq 0, \alpha \geq 0, \beta \geq 0, \quad (2.6.12)$$

where $\Gamma(\alpha)$ is the Gamma function. The Gamma pdf has two parameters α and β. The corresponding cumulative distribution function (cdf) to this pdf is denoted by $G(\alpha, \beta)$. When we write $v \sim G(\alpha, \beta)$, we mean that v is a random variable with pdf given in (2.6.12). The mgf corresponding to this distribution is expressed by

$$M_v(u) = (1 + \beta u)^{-\alpha}. \quad (2.6.13)$$

Recall from (2.6.1) and (2.6.2) that the reference window samples X_i can be described as

$$X_i \sim G(1, 2\sigma), \quad i = 1, 2, ..., N. \quad (2.6.14)$$

Also, Z is described as

$$Z \sim G(N, 2\sigma). \quad (2.6.15)$$

Using (2.6.8), P_F in a homogeneous background can be expressed as

$$P_F = M_Z(T/2\sigma)$$

$$= \left(1 + 2\sigma \frac{T}{2\sigma}\right)^{-N}$$

$$= (1 + T)^{-N}. \tag{2.6.16}$$

The value of the scale factor T, for a given false alarm probability and reference window size N, can be calculated as

$$T = -1 + P_F^{(-1/N)}. \tag{2.6.17}$$

Replacing T by $T/(1 + S)$ in (2.6.16), we obtain the probability of detection for given values of P_F, SNR and N as

$$P_D = [1 + T/(1 + S)]^{-N}. \tag{2.6.18}$$

Order Statistics CFAR Processor

The CA-CFAR processor performs well in a homogeneous environment. In a nonhomogeneous environment, its performance degrades significantly. Two commonly observed situations that give rise to nonhomogeneity are clutter edges and multiple targets in the reference window. Clutter edge refers to an abrupt noise power level transition within the reference window. When there are one or more targets (other than the primary target) within the reference window, we have the multiple target situation. In a nonhomogeneous situation, robust procedures are employed to estimate the background noise power level in CFAR processors. Resulting CFAR processors exhibit some performance loss in a homogeneous environment as compared to the CA-CFAR processor but perform much better in a nonhomogeneous environment.

One such CFAR processor is based on order statistics and is known as the OS-CFAR. In the OS-CFAR processor, the reference window cell samples, X_i, $i = 1, \ldots, N$, are rank-ordered according to increasing magnitude. This rank-ordered sequence is denoted as

$$X_{(1)} \leq X_{(2)} \leq \cdots \leq X_{(N)}. \tag{2.6.19}$$

The subscripts within parentheses mean that the variables X_i, $i = 1, 2, \ldots, N$, are rank-ordered. $X_{(1)}$ denotes the minimum and $X_{(N)}$ the maximum value. From this sequence, the kth ordered value, $X_{(k)}$, is selected as the statistic Z. The statistic Z is again multiplied by a scale factor, and the decision is made by comparing the test cell sample Y with TZ.

To determine the performance of the OS-CFAR processor, we need to know the pdf of the random variable $X_{(k)}$. In the homogeneous background, the pdf $p_{x_{(k)}}(z)$ is given by [Dav81]

$$p_{x_{(k)}}(z) = k \binom{N}{k} [1 - F(z)]^{N-k} [F(z)]^{k-1} p(z), \qquad (2.6.20)$$

where $p(\cdot)$ and $F(\cdot)$ are the probability density function (pdf) and cumulative distribution function (cdf) for the observations X_i, $i = 1, 2, ..., N$, respectively. In a Gaussian environment, $p(\cdot)$ is the exponential pdf given by (2.6.1) with $\lambda = \sigma$, and $F(\cdot)$ is the corresponding cdf. Inserting $p(\cdot)$ and $F(\cdot)$ into (2.6.20),

$$p_{x_{(k)}}(z) = \frac{k}{2\sigma} \binom{N}{k} (e^{-\frac{z}{2\sigma}})^{N-k+1} (1 - e^{-\frac{z}{2\sigma}})^{k-1}. \qquad (2.6.21)$$

Using (2.6.8), the false alarm probability can be computed as

$$\begin{aligned} P_F &= \frac{k}{2\sigma} \binom{N}{k} \int_0^\infty (1 - e^{-\frac{z}{2\sigma}})^{k-1} (e^{-\frac{z}{2\sigma}})^{T+N-k+1} dz \\ &= k \binom{N}{k} \int_0^\infty (1 - e^{-z})^{k-1} (e^{-z})^{T+N-k+1} dz \\ &= \prod_{i=0}^{k-1} (N-i)/(N-i+T). \qquad (2.6.22) \end{aligned}$$

The value of the scale factor T for a given false alarm probability can be computed iteratively from (2.6.22) while the reference window size N and the order number k are held fixed.

Replacing T by $T/(1 + S)$ in (2.6.22), we obtain the probability of detection for given values of P_F, SNR, N and k as

$$P_D = \prod_{i=0}^{k-1} \frac{(N-i)}{N-i+\dfrac{T}{(1+S)}}. \qquad (2.6.23)$$

From (2.6.22) and (2.6.23) we can observe that the performance of the OS-CFAR processor is independent of σ for the exponential noise model. It must be noted that this independence is not true in general but depends on the type of noise distribution assumed.

Nonhomogeneous Background

The performance of the CA-CFAR and the OS-CFAR processors in a nonhomogeneous environment can be evaluated similarly. These details are not provided here and can be found in [GaK88, Une93]. Here, we have presented only two CFAR processors. Many other CFAR algorithms have been proposed in the literature. These algorithms are designed to address specific issues, such as clutter edges, interfering targets, and non-Gaussian clutter.

2.7 Locally Optimum Detection

In this section we consider weak signal detection problems where the signal strength is very small relative to the background disturbance. An example of this situation arises in radar signal detection when targets are immersed in heavy clutter. Consider the binary hypothesis testing problem

$$\begin{aligned} H_0 &: y = n, \\ H_1 &: y = \theta s + n, \end{aligned} \qquad (2.7.1)$$

where y is the received observation, s is a known signal, n is noise and θ is a positive real-valued parameter. For the weak signal detection problem under consideration here, θ is small. The problem is to determine the most powerful test, i.e., the one that maximizes P_D, at some given level α. The Neyman–Pearson test for this problem is given

2.7 Locally Optimum Detection

by (2.4.3), and the threshold that satisfies the constraint on P_F is given by (2.4.4). When the value of θ is very small and $\theta \to 0$, $p(y|H_1) \to p(y|H_0)$. In this case, the likelihood ratio tends to one, and it becomes difficult to discriminate between the two hypotheses. One needs to employ other performance measures for system optimization. A popular approach is to maximize the slope of the power function or $P_D(\theta)$ at $\theta = 0$ under the constraint $P_F = \alpha$. The resulting test is known as a locally optimum test. Next, we employ the Lagrange multiplier method to derive the nonrandomized locally optimum test.

Consider the function that partitions the observation space into the two decision regions

$$\delta(y) = \begin{cases} 1, & \text{if } H_1 \text{ is declared true}, \\ 0, & \text{if } H_0 \text{ is declared true}. \end{cases} \quad (2.7.2)$$

The probabilities of detection and false alarm can be expressed as

$$P_D = P(\delta(y) = 1 | H_1) = \int_Z \delta(y) p(y|H_1) dy, \quad (2.7.3)$$

and

$$P_F = P(\delta(y) = 1 | H_0) = \int_Z \delta(y) p(y|H_0) dy, \quad (2.7.4)$$

where Z is the entire observation space. The objective is to determine $\delta(y)$ so that the slope of $P_D(\theta)$ is maximized at $\theta = 0$ under the constraint $P_F = \alpha$, i.e., maximize

$$\left. \frac{\partial P_D(\theta)}{\partial \theta} \right|_{\theta=0} = \left[\frac{\partial}{\partial \theta} \int_Z \delta(y) p(y|H_1) dy \right]_{\theta=0} \quad (2.7.5)$$

under the constraint

$$\int_Z \delta(y) p(y|H_0) dy = \alpha. \qquad (2.7.6)$$

We assume that the density function $p(y|H_1)$ is a well-behaved function so that the operations of integration and differentiation can be interchanged. Then, the objective function to be maximized is

$$\begin{aligned} F &= \left[\int_Z \delta(y) \frac{\partial}{\partial \theta} p(y|H_1) dy \right]_{\theta=0} + \lambda \left[\alpha - \int_Z \delta(y) p(y|H_0) dy \right] \\ &= \left[\int_Z \delta(y) \left[\frac{\partial}{\partial \theta} p(y|H_1)|_{\theta=0} - \lambda p(y|H_0) \right] dy \right] + \lambda \alpha, \end{aligned} \qquad (2.7.7)$$

where λ is the unknown Lagrange multiplier. It is easy to see that the resulting decision rule is

$$\frac{\frac{\partial}{\partial \theta} p(y|H_1)|_{\theta=0}}{p(y|H_0)} \begin{array}{c} H_1 \\ > \\ < \\ H_0 \end{array} \lambda, \qquad (2.7.8)$$

where λ is determined from the constraint (2.7.6). If, for the problem under consideration,

$$\frac{\partial}{\partial \theta} p(y|H_1)|_{\theta=0} = 0, \qquad (2.7.9)$$

we need to select another performance measure for system optimization. In this case, the second derivative of the power function at $\theta = 0$ is maximized under the constraint $P_F = \alpha$. It can be easily shown that the resulting test is expressed by

$$\frac{\frac{\partial^2}{\partial \theta^2} p(y|H_1)|_{\theta=0}}{p(y|H_0)} \overset{H_1}{\underset{H_0}{\gtrless}} \lambda, \qquad (2.7.10)$$

where λ is again obtained from (2.7.6).

In the above development only nonrandomized tests were considered. A more general formulation that allows randomized tests, similar to the one shown in (2.4.5), can also be considered.

Notes and Suggested Reading

One can find basic treatment of detection theory in a wide variety of books. Examples are [Bar91], [Hel95a], [MeC78], [Poo88], [SRV95] and [Van68]. Details on CA-CFAR and OS-CFAR can be found in [FiJ68] and [Roh83] respectively. Other recent results on CFAR are available in papers published in *IEEE Transactions on Aerospace and Electronic Systems*. A more detailed treatment of locally optimum detection is available in [Mid60] and [Kas88].

3
Distributed Bayesian Detection: Parallel Fusion Network

3.1 Introduction

As indicated earlier, detection networks can be organized in a number of topological structures. Among the topologies considered in the literature, the parallel fusion topology has received the most attention. In this chapter, we develop the theory of Bayesian detection for parallel fusion structures. In Section 3.2, we consider a parallel structure consisting of a number of detectors whose decisions are available locally and are not transmitted to a fusion center for decision combining. Costs of decision making are assumed to be coupled and a system wide optimization is carried out for binary and ternary hypothesis testing problems. Section 3.3 considers the design of fusion rules given the statistics of incoming decisions. Design of the parallel fusion network, consisting of a number of local detectors and a fusion center, is the subject of Section 3.4. Person-by-person optimal decision rules are derived. A number of special cases including conditionally independent local observations and identical detectors are considered. Efficient computational approaches are presented. Design of optimal parallel structures with dependent local observations is an NP-complete problem. This and other computational complexity issues are briefly considered. Finally, robust detection and nonparametric detection are discussed at the end of the chapter.

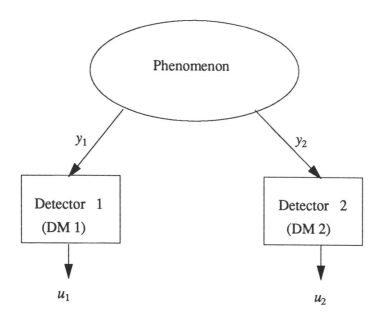

Figure 3.1. A two detector parallel fusion network without fusion.

3.2 Distributed Detection Without Fusion

We begin by considering the parallel detection network without fusion. This topology is shown in Figure 3.1 for the case of two detectors or decision makers DM 1 and DM 2. In this case, both detectors observe a common phenomenon and make local decisions regarding it. These local decisions are not fused to obtain a global decision. The detectors do not communicate with each other but, their operation is coupled due to the fact that the costs of decision making are coupled and a system-wide optimization is performed. We first consider the binary hypothesis testing problem. Let H_0 and H_1 be the two hypotheses with P_0 and P_1 denoting the associated a priori probabilities. The observations at the two detectors are denoted by y_1 and y_2, respectively. The joint conditional density functions under the two hypotheses are $p(y_1, y_2|H_i)$, $i = 0, 1$. The decisions at the two detectors are denoted by u_i, $i = 1, 2$, and are given by

$$u_i = \begin{cases} 0, & H_0 \text{ is declared present,} \\ 1, & H_1 \text{ is declared present.} \end{cases} \quad (3.2.1)$$

The local decisions u_i are based only on their local observations y_i, i.e., there is no communication between the detectors. The costs of different courses of action are denoted by C_{ijk}, $i, j, k = 0, 1$, where C_{ijk} represents the cost of detector 1 deciding H_i, detector 2 deciding H_j when H_k is present. The objective is to obtain decision rules at the two detectors that minimize the average cost of decision making. The Bayes risk function is given by

$$\begin{aligned} \Re &= \sum_{i,j,k} \int_{y_1,y_2} p(u_1,u_2,y_1,y_2,H_k) C_{ijk} \\ &= \sum_{i,j,k} \int_{y_1,y_2} P_k C_{ijk} p(u_1,u_2,y_1,y_2|H_k) \\ &= \sum_{i,j,k} \int_{y_1,y_2} P_k C_{ijk} p(u_1,u_2|y_1,y_2,H_k) p(y_1,y_2|H_k) \:. \end{aligned} \quad (3.2.2)$$

Note that the local decisions u_1 and u_2 are independent and are made based only on their local observations y_1 and y_2, respectively. Furthermore, they do not depend upon the hypothesis present. Therefore,

$$\Re = \sum_{i,j,k} \int_{y_1,y_2} P_k C_{ijk} p(u_1|y_1) p(u_2|y_2) p(y_1,y_2|H_k). \quad (3.2.3)$$

We expand the right-hand side by explicitly summing over u_1 and get

$$\begin{aligned} \Re = \sum_{j,k} \int_{y_1,y_2} &P_k \, p(u_2|y_2) p(y_1,y_2|H_k) \\ &\times \left[C_{0jk} p(u_1=0|y_1) + C_{1jk} p(u_1=1|y_1) \right]. \end{aligned} \quad (3.2.4)$$

Because $p(u_1=1|y_1) = 1 - p(u_1=0|y_1)$, we express $\Re$ as

$$\mathfrak{R} = \sum_{j,k} \int_{y_1,y_2} P_k p(u_2|y_2) p(y_1,y_2|H_k)$$
$$\times \left[C_{0jk} p(u_1=0|y_1) + C_{1jk}(1 - p(u_1=0|y_1)) \right]. \quad (3.2.5)$$

Ignoring the following constant term with respect to u_1,

$$\sum_{j,k} \int_{y_1,y_2} C_{1jk} P_k p(u_2|y_2) p(y_1,y_2|H_k),$$

$\mathfrak{R}$ becomes

$$\mathfrak{R} = \int_{y_1} p(u_1=0|y_1) \sum_{j,k} \int_{y_2} P_k p(u_2|y_2) p(y_1,y_2|H_k) [C_{0jk} - C_{1jk}].$$
$$(3.2.6)$$

$\mathfrak{R}$ is minimized if we set

$$p(u_1=0|y_1) = \begin{cases} 0, & \text{if } \sum_{j,k} \int_{y_2} P_k p(u_2|y_2) p(y_1,y_2|H_k)[C_{0jk} - C_{1jk}] \geq 0 \\ 1, & \text{otherwise} . \end{cases} \quad (3.2.7)$$

Note that the decision rule at detector 1 is a deterministic rule (not a randomized rule) regardless of the costs, decision rule at the other detector, and the joint conditional density of the observations. Therefore, we can express (3.2.7) as the decision rule

$$\sum_{j,k} \int_{y_2} P_k p(u_2|y_2) p(y_1,y_2|H_k)[C_{0jk} - C_{1jk}] \underset{u_1=0}{\overset{u_1=1}{\gtrless}} 0. \quad (3.2.8)$$

Now expanding the sum over k,

$$\sum_j \int_{y_2} P_0 p(u_2|y_2) p(y_1,y_2|H_0) [C_{0j0} - C_{1j0}]$$

$$+ \sum_j \int_{y_2} P_1 p(u_2|y_2) p(y_1,y_2|H_1) [C_{0j1} - C_{1j1}] \overset{u_1=1}{\underset{u_1=0}{\gtrless}} 0. \qquad (3.2.9)$$

Taking the second sum to the right-hand side,

$$\sum_j \int_{y_2} P_0 p(u_2|y_2) p(y_1,y_2|H_0) [C_{0j0} - C_{1j0}] \overset{u_1=1}{\underset{u_1=0}{\gtrless}}$$

$$- \sum_j \int_{y_2} P_1 p(u_2|y_2) p(y_1,y_2|H_1) [C_{0j1} - C_{1j1}]. \qquad (3.2.10)$$

We make the assumption that $C_{0j0} < C_{1j0}$, i.e., the cost of detector 1 making an error when H_0 is present is more than the cost of it being right regardless of the decision of detector 2. Also,

$$p(y_1,y_2|H_k) = p(y_2|y_1,H_k) p(y_1|H_k), \quad k = 0, 1. \qquad (3.2.11)$$

Then, (3.2.10) can be expressed as

$$- \sum_j \int_{y_2} P_0 p(u_2|y_2) p(y_2|y_1,H_0) p(y_1|H_0) [C_{1j0} - C_{0j0}] \overset{u_1=1}{\underset{u_1=0}{\gtrless}}$$

$$- \sum_j \int_{y_2} P_1 p(u_2|y_2) p(y_2|y_1,H_1) p(y_1|H_1) [C_{0j1} - C_{1j1}]. \qquad (3.2.12)$$

Taking the terms not dependent on j and y_2 outside and rearranging terms,

3.2 Distributed Detection Without Fusion

$$-P_0 p(y_1|H_0) \sum_j \int_{y_2} p(u_2|y_2) p(y_2|y_1,H_0)[C_{1j0} - C_{0j0}] \underset{u_1=0}{\overset{u_1=1}{\gtrless}}$$

$$-P_1 p(y_1|H_1) \sum_j \int_{y_2} p(u_2|y_2) p(y_2|y_1,H_1)[C_{0j1} - C_{1j1}],$$

or

$$\frac{P_1 p(y_1|H_1)}{P_0 p(y_1|H_0)} \underset{u_1=0}{\overset{u_1=1}{\gtrless}} \frac{\sum_j \int_{y_2} p(u_2|y_2) p(y_2|y_1,H_0)[C_{1j0} - C_{0j0}]}{\sum_j \int_{y_2} p(u_2|y_2) p(y_2|y_1,H_1)[C_{0j1} - C_{1j1}]},$$

or

$$\Lambda(y_1) \underset{u_1=0}{\overset{u_1=1}{\gtrless}} \frac{P_0 \sum_j \int_{y_2} p(u_2|y_2) p(y_2|y_1,H_0)[C_{1j0} - C_{0j0}]}{P_1 \sum_j \int_{y_2} p(u_2|y_2) p(y_2|y_1,H_1)[C_{0j1} - C_{1j1}]},$$

(3.2.13)

where $\Lambda(y_1)$ represents the likelihood ratio at detector 1 and is given by

$$\Lambda(y_1) = \frac{p(y_1|H_1)}{p(y_1|H_0)}.$$

Note that the right-hand side (RHS) of (3.2.13) is not a simple constant threshold. It is data dependent, i.e., it depends on y_1 due to the appearance of the term $p(y_2|y_1,H_k)$, $k = 0, 1$. Also, the right-hand side is a function of the decision rule at the other detector due to the term $p(u_2|y_2)$.

The situation simplifies considerably if y_1 and y_2 are assumed to be conditionally independent, given any hypothesis. In this case, the right-hand side of (3.2.13) is no longer data dependent. Under this condition,

$$p(y_2|y_1,H_k) = p(y_2|H_k), \quad k = 0, 1, \qquad (3.2.14)$$

and the right-hand side of (3.2.13) reduces to a threshold t_1 which still depends on the decision rule of the other detector, i.e., its threshold t_2. The threshold t_1 is given by

$$t_1 = \frac{P_0 \sum_j \int_{y_2} p(u_2|y_2) p(y_2|H_0) [C_{1j0} - C_{0j0}]}{P_1 \sum_j \int_{y_2} p(u_2|y_2) p(y_2|H_1) [C_{0j1} - C_{1j1}]}. \qquad (3.2.15)$$

Expanding over j, we get

$$t_1 = \frac{P_0 \int_{y_2} p(y_2|H_0) \{p(u_2=0|y_2)[C_{100}-C_{000}] + p(u_2=1|y_2)[C_{110}-C_{010}]\}}{P_1 \int_{y_2} p(y_2|H_1) \{p(u_2=0|y_2)[C_{001}-C_{101}] + p(u_2=1|y_2)[C_{011}-C_{111}]\}}.$$

$$(3.2.16)$$

Noting that $p(u_2 = 1|y_2) = 1 - p(u_2 = 0|y_2)$ and rearranging terms, t_1 becomes

$$t_1 = \frac{P_0 \int_{y_2} p(y_2|H_0)\{[C_{110}-C_{010}] + p(u_2=0|y_2)[C_{100}-C_{000}+C_{010}-C_{110}]\}}{P_1 \int_{y_2} p(y_2|H_1)\{[C_{011}-C_{111}] + p(u_2=0|y_2)[C_{001}-C_{101}+C_{111}-C_{011}]\}}.$$

$$(3.2.17)$$

The threshold t_1 is a function of $p(u_2 = 0|y_2)$ which specifies the decision rule at the second detector, and, thus, it is a function of t_2. Therefore,

$$t_1 = f_1(t_2), \qquad (3.2.18)$$

where $f_1(\cdot)$ is the function defined in (3.2.17). Similarly,

$$t_2 = f_2(t_1), \qquad (3.2.19)$$

where $f_2(\cdot)$ has a form similar to $f_1(\cdot)$. Equations (3.2.18) and (3.2.19) specify necessary conditions that the thresholds t_1 and t_2 must satisfy if they are to be optimal. These conditions are not sufficient. Their solution provides locally optimum solutions. When there are several local minima, each must be examined to determine the globally optimum solution. It is important to observe that the two thresholds are coupled and this is a result of our objective of systemwide optimization. In general, the resulting thresholds are not the same as would be obtained if each local detector was optimized independently.

Next, we consider a special case where the thresholds t_1 and t_2 decouple. Let the cost assignment be

$$C_{000} = C_{111} = 0,$$

$$C_{010} = C_{100} = C_{011} = C_{101} = 1,$$

$$C_{110} = C_{001} = k. \qquad (3.2.20)$$

This cost function is symmetric in u_1 and u_2. All cases of equal number of errors are assigned equal costs. Let us assume here, $k = 2$. Substituting these costs in (3.2.17),

$$\begin{aligned} t_1 &= \frac{P_0 \int_{y_2} p(y_2|H_0)[1 + p(u_2 = 0|y_2) \times 0]}{P_1 \int_{y_2} p(y_2|H_1)[1 + p(u_2 = 0|y_2) \times 0]} \\ &= \frac{P_0 \int_{y_2} p(y_2|H_0)}{P_1 \int_{y_2} p(y_2|H_1)} \\ &= \frac{P_0}{P_1}. \end{aligned} \qquad (3.2.21)$$

In this special case, the threshold t_1 is a constant and is the same as obtained by independently applying the minimum probability of error decision rule at each local detector. It should be emphasized that the cost assignment (3.2.20) with $k = 2$ is a very special cost assignment where t_1 and t_2 are independent of each other. This is not the case in general. Consider the following examples for illustration.

Example 3.1

Let us assume that the two local observations are conditionally independent, identical and are Gaussian distributed. The conditional densities are given by

$$p(y_i|H_j) = \frac{1}{\sigma\sqrt{2\pi}} \exp\left(-\frac{(y_i-m_j)^2}{2\sigma^2}\right), \quad i = 1, 2; \quad j = 0, 1; -\infty < y_i < \infty.$$

(3.2.22)

Assume that the a priori probabilities P_0 and P_1 are equal. The likelihood ratio at the two detectors is given by

$$\Lambda(y_i) = \frac{p(y_i|H_1)}{p(y_i|H_0)}$$

$$= \exp\left(\frac{1}{2\sigma^2}[2y_i(m_1 - m_0) - (m_1^2 - m_0^2)]\right), \quad i = 1, 2.$$

(3.2.23)

The likelihood ratio test at each detector is given by

$$\Lambda(y_i) \underset{u_i = 0}{\overset{u_i = 1}{\gtrless}} t_i,$$

(3.2.24)

which reduces to

$$y_i \underset{u_i = 0}{\overset{u_i = 1}{\gtrless}} \frac{2\sigma^2 \log t_i + (m_1^2 - m_0^2)}{2(m_1 - m_0)}, \quad i = 1, 2.$$

(3.2.25)

3.2 Distributed Detection Without Fusion

Let us assume that the cost assignment is given by (3.2.20). The thresholds t_1 and t_2 can be obtained from (3.2.17) and its companion equation. The threshold t_1 is given by

$$t_1 = \frac{P_0 \int_{y_2} p(y_2|H_0)\{(k-1) + p(u_2=0|y_2)[2-k]\}}{P_1 \int_{y_2} p(y_2|H_1)\{1 + p(u_2=0|y_2)[k-2]\}}$$

$$= \frac{(k-1)\int_{y_2} p(y_2|H_0) + (2-k)\int_{y_2} p(u_2=0|y_2)p(y_2|H_0)}{\int_{y_2} p(y_2|H_1) + (k-2)\int_{y_2} p(u_2=0|y_2)p(y_2|H_1)}. \quad (3.2.26)$$

Because

$$p(u_2=0|y_2) = p(u_2=0|y_2, H_j), \quad j=0,1,$$

$$\int_{y_2} p(u_2=0|y_2)p(y_2|H_j) = \int_{y_2} p(u_2=0|y_2, H_j)p(y_2|H_j)$$

$$= p(u_2=0|H_j), \quad j=0,1. \quad (3.2.27)$$

Also,

$$\int_{y_2} p(y_2|H_j) = 1, \quad j=0,1.$$

Therefore,

$$t_1 = \frac{(k-1) + (2-k)p(u_2=0|H_0)}{1 + (k-2)p(u_2=0|H_1)}. \quad (3.2.28)$$

For Gaussian observation statistics, the probabilities $p(u_2=0|H_j)$, $j=0,1$, can be easily obtained to be

3. Distributed Bayesian Detection: Parallel Fusion Network

$$p(u_2 = 0 | H_0) = \text{erf}\left(\frac{\sigma \log t_2}{m_1 - m_0} + \frac{m_1 - m_0}{2\sigma}\right),$$

and

$$p(u_2 = 0 | H_1) = \text{erf}\left(\frac{\sigma \log t_2}{m_1 - m_0} - \frac{m_1 - m_0}{2\sigma}\right),$$

where

$$\text{erf}(x) \triangleq \int_{-\infty}^{x} \frac{1}{\sqrt{2\pi}} \exp\left(-\frac{y^2}{2}\right) dy.$$

Substituting these probabilities into (3.2.28) and its companion equation,

$$t_1 = \frac{(k-1) + (2-k) \, \text{erf}\left(\frac{\sigma \log t_2}{m_1 - m_0} + \frac{m_1 - m_0}{2\sigma}\right)}{1 + (k-2) \, \text{erf}\left(\frac{\sigma \log t_2}{m_1 - m_0} - \frac{m_1 - m_0}{2\sigma}\right)}, \quad (3.2.29)$$

and

$$t_2 = \frac{(k-1) + (2-k) \, \text{erf}\left(\frac{\sigma \log t_1}{m_1 - m_0} + \frac{m_1 - m_0}{2\sigma}\right)}{1 + (k-2) \, \text{erf}\left(\frac{\sigma \log t_1}{m_1 - m_0} - \frac{m_1 - m_0}{2\sigma}\right)}. \quad (3.2.30)$$

The solution of (3.2.29) and (3.2.30) yields the thresholds t_1 and t_2. As noted before, multiple solutions may result, and each one must be examined to determine the best solution. Consider the solution of (3.2.29) and (3.2.30) for different values of k. Assume that $m_0 = 0$, $m_1 = 1$, and $\sigma = 1$. For $1 \leq k < 4.528$, there is only one solution $t_1 = t_2 = 1$ that yields the minimum value of $\Re$. For $k \geq 4.528$, there are three solutions. One of the solutions is $t_1 = t_2 = 1$ but it does not yield the minimum value of $\Re$. The other two solutions need to be used in a pair, i.e., one threshold is set equal to one solution and the other threshold is set equal to the other solution. This setting of unequal

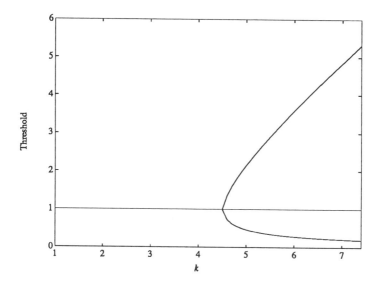

Figure 3.2. Threshold values as a function of k for Example 3.1.

thresholds yields the minimum value of $\mathfrak{R}$. Threshold values as a function of k are shown in Figure 3.2. Note that an interesting situation has arisen in this example. Even though observations at each detector are identically distributed and the cost function is symmetric, unequal thresholds are optimal for $k \geq 4.528$. This is due to the large cost assigned to double errors, and the optimum action on the part of the detectors is to employ different thresholds.

Example 3.2

Consider again the case of two conditionally independent and identically distributed local observations. The conditional densities are assumed to be Rayleigh and are given by

$$p(y_i|H_j) = \frac{y_i}{\sigma_j^2} \exp\left(-\frac{y_i^2}{2\sigma_j^2}\right), \quad i = 1, 2; \quad j = 0, 1; \quad y_i \geq 0. \quad (3.2.31)$$

The likelihood ratio at the two detectors is

$$\Lambda(y_i) = \frac{\sigma_0^2}{\sigma_1^2} \exp\left[\frac{1}{2}\left(\frac{1}{\sigma_0^2} - \frac{1}{\sigma_1^2}\right) y_i^2\right], \quad i = 1, 2. \qquad (3.2.32)$$

The likelihood ratio test at each detector

$$\Lambda(y_i) \underset{u_i=0}{\overset{u_i=1}{\gtrless}} t_i, \quad i = 1, 2,$$

reduces to

$$y_i \underset{u_i=0}{\overset{u_i=1}{\gtrless}} \left[\frac{2\sigma_0^2 \sigma_1^2}{\sigma_1^2 - \sigma_0^2} \log\left(\frac{\sigma_1^2 t_i}{\sigma_0^2}\right)\right]^{1/2} \triangleq \psi(t_i). \qquad (3.2.33)$$

Assuming equal a priori probabilities and the cost assignment (3.2.20), the thresholds are as given in (3.2.28). In this example, the probabilities $p(u_2 = 0|H_j)$, $j = 1, 2$, are

$$p(u_2 = 0|H_j) = \int_0^{\psi(t_2)} \frac{y_2}{\sigma_j^2} \exp\left(-\frac{y_2^2}{2\sigma_j^2}\right) dy_2$$

$$= 1 - \exp\left(-\frac{\psi^2(t_2)}{2\sigma_j^2}\right), \quad j = 0, 1. \qquad (3.2.34)$$

Substituting these in (3.2.28) and simplifying,

$$t_1 = \frac{1 + (k-2) \exp\left(-\frac{\psi^2(t_2)}{2\sigma_0^2}\right)}{(k-1) - (k-2) \exp\left(-\frac{\psi^2(t_2)}{2\sigma_1^2}\right)}, \qquad (3.2.35)$$

and

$$t_2 = \frac{1 + (k-2)\exp\left(-\frac{\psi^2(t_1)}{2\sigma_0^2}\right)}{(k-1) - (k-2)\exp\left(-\frac{\psi^2(t_1)}{2\sigma_1^2}\right)}. \qquad (3.2.36)$$

Equations (3.2.35) and (3.2.36) represent the necessary conditions that the thresholds must satisfy. As before, we compute the thresholds for different values of k and examine the multiplicity of solutions. We assume $\sigma_0 = 1$ and $\sigma_1 = 2$. Because $y_i \geq 0$ for Rayleigh distribution, we conclude from (3.2.32) that $\Lambda(y_i) \geq \sigma_0^2/\sigma_1^2$ and, therefore, the minimum value of t_i in this example is 1/4. For any solution $t_i < 1/4$, t_i can be set equal to 1/4. The solutions for different values of k are sketched in Figure 3.3. For $1 \leq k < 2$, there are two solutions, and each solution corresponds to a setting of $t_1 = t_2$. For $1 \leq k \leq 1.25$, both solutions are greater than 1/4. But when $1.25 < k < 2$, one of the solutions is less than 1/4 and we set it equal to 1/4 as shown in Figure 3.3. For $2 \leq k < 5.98$, there is a single solution requiring a $t_1 = t_2$ setting. For $k \geq 5.98$, there are three solutions, one of which is the $t_1 = t_2$ solution. For $5.98 \leq k < 7.05$, both of the unequal threshold solutions are greater than 1/4. For k

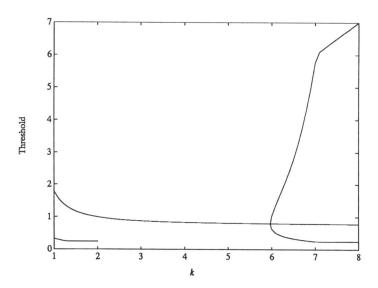

Figure 3.3. Threshold values as a function of k for Example 3.2.

≥ 7.05, one of the solutions becomes less than 1/4 and is, therefore, set equal to 1/4. The other one is determined based on the threshold value 1/4. As far as the optimality is concerned, for $1 \leq k < 5.98$, the upper continuous $t_1 = t_2$ solution shown in Figure 3.3 yields the minimum risk. For $k \geq 5.98$, the $t_1 \neq t_2$ solution results in a smaller value of risk and is, therefore, used.

M-ary Hypothesis Testing Using N Sensors

Next, we consider the more general M-hypothesis, N-sensor distributed detection problem. The system topology is shown in Figure 3.4. Once again, N detectors observe a common phenomenon and make local decisions regarding the hypothesis present. The local decisions are not combined. Let $H_0, H_1, ..., H_{M-1}$ denote the M hypotheses with a priori probabilities $P_0, P_1, ..., P_{M-1}$. For M hypotheses, there are M^N possible sequences of local decisions. Therefore, there are M^{N+1} alternatives that may occur each time the hypothesis testing task is carried out. For ease of presentation and notational simplicity, we confine our attention to the $M = 3$ and $N = 2$ case in this book, i.e., we only consider the three hypothesis problem using two detectors. Local observations are again denoted by y_1 and y_2 with the joint conditional density $p(y_1, y_2|H_i)$, $i = 0, 1, 2$. The local decisions u_i, $i = 1, 2$, are given by

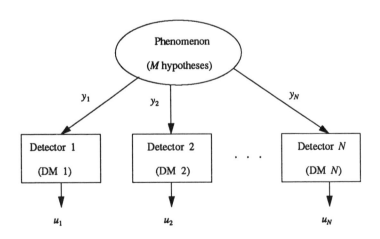

Figure 3.4. An N-detector parallel fusion network without fusion.

3.2 Distributed Detection Without Fusion 51

$$u_i = \begin{cases} 0, & H_0 \text{ is declared present,} \\ 1, & H_1 \text{ is declared present,} \\ 2, & H_2 \text{ is declared present.} \end{cases} \quad (3.2.37)$$

The cost of detector 1 deciding H_i, detector 2 deciding H_j, when H_k is present, is represented by C_{ijk}, $i,j,k = 0, 1, 2$. Local decision rules are to be obtained that minimize the Bayes risk function $\Re$ given by

$$\Re = \sum_{i,j,k} \int_{y_1,y_2} p(u_1,u_2,y_1,y_2,H_k) C_{ijk}$$

$$= \sum_{i,j,k} \int_{y_1,y_2} P_k C_{ijk} p(u_1,u_2|y_1,y_2,H_k) p(y_1,y_2|H_k). \quad (3.2.38)$$

Because the local decisions are independent, depending only on their observations and not on the hypothesis present, $\Re$ may be expressed as

$$\Re = \sum_{i,j,k} \int_{y_1,y_2} P_k C_{ijk} p(u_1|y_1) p(u_2|y_2) p(y_1,y_2|H_k). \quad (3.2.39)$$

Expanding by explicitly summing over u_1,

$$\Re = \int_{y_1} p(u_1=0|y_1) \sum_{j,k} \int_{y_2} P_k p(u_2|y_2) p(y_1,y_2|H_k) C_{0jk}$$
$$+ \int_{y_1} p(u_1=1|y_1) \sum_{j,k} \int_{y_2} P_k p(u_2|y_2) p(y_1,y_2|H_k) C_{1jk}$$
$$+ \int_{y_1} p(u_1=2|y_1) \sum_{j,k} \int_{y_2} P_k p(u_2|y_2) p(y_1,y_2|H_k) C_{2jk}.$$

(3.2.40)

Let Z_1 denote the entire observation space of detector 1 and let Z_{10}, Z_{11} and Z_{12} be the decision regions of detector 1 corresponding to H_0, H_1 and H_2, respectively. Then, $p(u_1 = i|y_1) = 1$, if $y_1 \in Z_{1i}$. The Bayes risk $\Re$ can be expressed as

$$\Re = \int_{Z_{10}} \sum_{j,k} \int_{y_2} P_k p(u_2|y_2) p(y_1,y_2|H_k) C_{0jk}$$

$$+ \int_{Z_{11}} \sum_{j,k} \int_{y_2} P_k p(u_2|y_2) p(y_1,y_2|H_k) C_{1jk}$$

$$+ \int_{Z_{12}} \sum_{j,k} \int_{y_2} P_k p(u_2|y_2) p(y_1,y_2|H_k) C_{2jk}$$

$$= \int_{Z_{10}} \sum_{j} \int_{y_2} p(u_2|y_2) [P_0 C_{0j0} p(y_1,y_2|H_0)$$

$$+ P_1 C_{0j1} p(y_1,y_2|H_1) + P_2 C_{0j2} p(y_1,y_2|H_2)]$$

$$+ \int_{Z_{11}} \sum_{j} \int_{y_2} p(u_2|y_2) [P_0 C_{1j0} p(y_1,y_2|H_0)$$

$$+ P_1 C_{1j1} p(y_1,y_2|H_1) + P_2 C_{1j2} p(y_1,y_2|H_2)]$$

$$+ \int_{Z_{12}} \sum_{j} \int_{y_2} p(u_2|y_2) [P_0 C_{2j0} p(y_1,y_2|H_0)$$

$$+ P_1 C_{2j1} p(y_1,y_2|H_1) + P_2 C_{2j2} p(y_1,y_2|H_2)]. \quad (3.2.41)$$

Using the fact that $Z_1 = Z_{10} + Z_{11} + Z_{12}$, leaving out the constant term involving the integration over the entire observation space Z_1, and rearranging,

$$\Re = \int_{Z_{10}} \sum_{j} \int_{y_2} p(u_2|y_2) [P_1(C_{0j1} - C_{1j1}) p(y_1,y_2|H_1)$$

$$+ P_2(C_{0j2} - C_{2j2}) p(y_1,y_2|H_2)]$$

$$+ \int_{Z_{11}} \sum_{j} \int_{y_2} p(u_2|y_2) [P_0(C_{1j0} - C_{0j0}) p(y_1,y_2|H_0)$$

$$+ P_2(C_{1j2} - C_{2j2}) p(y_1,y_2|H_2)]$$

$$+ \int_{Z_{12}} \sum_{j} \int_{y_2} p(u_2|y_2) [P_0(C_{2j0} - C_{0j0}) p(y_1,y_2|H_0)$$

$$+ P_1(C_{2j1} - C_{1j1}) p(y_1,y_2|H_1)]$$

$$= \int_{Z_{10}} I'_{10}(y_1) + \int_{Z_{11}} I'_{11}(y_1) + \int_{Z_{12}} I'_{12}(y_1), \quad (3.2.42)$$

where $I'_{10}(\cdot)$, $I'_{11}(\cdot)$ and $I'_{12}(\cdot)$ represent the integrands of (3.2.42). To minimize $\Re$, the following decision rule at detector 1 should be used.

$$p(u_1 = 0|y_1) = \begin{cases} 1, & \text{if } I'_{10}(y_1) < \min\left(I'_{11}(y_1), I'_{12}(y_1)\right), \\ 0, & \text{otherwise}, \end{cases} \quad (3.2.43a)$$

$$p(u_1 = 1|y_1) = \begin{cases} 1, & \text{if } I'_{11}(y_1) < \min\left(I'_{10}(y_1), I'_{12}(y_1)\right), \\ 0, & \text{otherwise}, \end{cases} \quad (3.2.43b)$$

and

$$p(u_1 = 2|y_1) = \begin{cases} 1, & \text{if } I'_{12}(y_1) < \min\left(I'_{10}(y_1), I'_{11}(y_1)\right), \\ 0, & \text{otherwise}. \end{cases} \quad (3.2.43c)$$

Detector 2 employs a similar decision rule. A simultaneous solution of the decision rule inequalities yields the observation space partitions at the two local detectors.

Next, we assume conditional independence of observations at the two sensors, i.e.,

$$p(y_1, y_2 | H_k) = p(y_1 | H_k) p(y_2 | H_k), \quad k = 0, 1, 2. \quad (3.2.44)$$

We also make the assumption that the cost of making a wrong decision by a sensor is higher than the cost of making a correct decision regardless of the decision of the other sensor. This means that all the cost differences appearing in (3.2.42) are positive. Let $\Lambda_{11}(y_1)$ and $\Lambda_{12}(y_1)$ be likelihood ratios defined at detector 1 as follows

$$\Lambda_{11}(y_1) = \frac{p(y_1 | H_1)}{p(y_1 | H_0)}, \quad (3.2.45a)$$

and

$$\Lambda_{12}(y_1) = \frac{p(y_1|H_2)}{p(y_1|H_0)}. \qquad (3.2.45b)$$

Dividing the integrands $I'_{10}(\cdot)$, $I'_{11}(\cdot)$, and $I'_{12}(\cdot)$ by $p(y_1|H_0)$, the resulting expressions in terms of the likelihood ratios Λ_{11} and Λ_{12} are as follows:

$$\begin{aligned} I_{10}(y_1) &= I'_{10}(y_1)/p(y_1|H_0) \\ &= \sum_j \int_{y_2} p(u_2|y_2)[P_1(C_{0j1} - C_{1j1})\Lambda_{11} p(y_2|H_1) \\ &\quad + P_2(C_{0j2} - C_{2j2})\Lambda_{12} p(y_2|H_2)], \end{aligned} \qquad (3.2.46)$$

$$\begin{aligned} I_{11}(y_1) &= \sum_j \int_{y_2} p(u_2|y_2)[P_0(C_{1j0} - C_{0j0}) p(y_2|H_0) \\ &\quad + P_2(C_{1j2} - C_{2j2})\Lambda_{12} p(y_2|H_2)], \end{aligned} \qquad (3.2.47)$$

and

$$\begin{aligned} I_{12}(y_1) &= \sum_j \int_{y_2} p(u_2|y_2)[P_0(C_{2j0} - C_{0j0}) p(y_2|H_0) \\ &\quad + P_1(C_{2j1} - C_{1j1})\Lambda_{11} p(y_2|H_1)]. \end{aligned} \qquad (3.2.48)$$

The decision rule at detector 1 is, then, given by (3.2.43), where $I_{10}(y_1)$, $I_{11}(y_1)$ and $I_{12}(y_1)$ are used instead of $I'_{10}(y_1)$, $I'_{11}(y_1)$ and $I'_{12}(y_1)$. This decision rule, in terms of the likelihood ratios defined in (3.2.45), is given below:

$$I_{10}(y_1) \underset{H_0 \text{ or } H_2}{\overset{H_1 \text{ or } H_2}{\gtrless}} I_{11}(y_1), \qquad (3.2.49a)$$

$$I_{11}(y_1) \underset{H_0 \text{ or } H_1}{\overset{H_0 \text{ or } H_2}{\gtrless}} I_{12}(y_1), \qquad (3.2.49b)$$

and

$$I_{10}(y_1) \underset{H_0 \text{ or } H_1}{\overset{H_1 \text{ or } H_2}{\gtrless}} I_{12}(y_1). \qquad (3.2.49c)$$

Let us now concentrate on $I_{10}(y_1)$ and expand it by summing it explicitly over u_2.

$$\begin{aligned}
I_{10}(y_1) = &\int_{y_2} p(u_2=0|y_2)[P_1(C_{001}-C_{101})\Lambda_{11}p(y_2|H_1) \\
&+P_2(C_{002}-C_{202})\Lambda_{12}p(y_2|H_2)] \\
&+\int_{y_2} p(u_2=1|y_2)[P_1(C_{011}-C_{111})\Lambda_{11}p(y_2|H_1) \\
&+P_2(C_{012}-C_{212})\Lambda_{12}p(y_2|H_2)] \\
&+\int_{y_2} p(u_2=2|y_2)[P_1(C_{021}-C_{121})\Lambda_{11}p(y_2|H_1) \\
&+P_2(C_{022}-C_{222})\Lambda_{12}p(y_2|H_2)]. \qquad (3.2.50)
\end{aligned}$$

Consider the following cost assignment for the three hypothesis problem. This assignment is similar to that given in (3.2.20) for the binary case. When any hypothesis is mistaken to be any of the other two hypotheses, the error is assumed to be equally costly. Also, all cases of equal number of errors are assumed to be equally costly.

$$C_{000} = C_{111} = C_{222} = 0,$$

$$C_{010} = C_{020} = C_{100} = C_{200} = 1,$$

$$C_{011} = C_{101} = C_{121} = C_{211} = 1,$$

$$C_{022} = C_{202} = C_{122} = C_{212} = 1,$$

$$C_{110} = C_{120} = C_{210} = C_{220} = k,$$

$$C_{001} = C_{021} = C_{201} = C_{221} = k,$$

$$C_{002} = C_{012} = C_{102} = C_{112} = k. \qquad (3.2.51)$$

Using this cost assignment, $I_{10}(y_1)$ becomes

$$I_{10}(y_1) = \int_{y_2} p(u_2=0|y_2)[P_1(k-1)\Lambda_{11}p(y_2|H_1) + P_2(k-1)\Lambda_{12}p(y_2|H_2)]$$
$$+ \int_{y_2} p(u_2=1|y_2)[P_1\Lambda_{11}p(y_2|H_1) + P_2(k-1)\Lambda_{12}p(y_2|H_2)]$$
$$+ \int_{y_2} p(u_2=2|y_2)[P_1(k-1)\Lambda_{11}p(y_2|H_1) + P_2\Lambda_{12}p(y_2|H_2)].$$

(3.2.52)

Because

$$p(u_2=j|y_2) = \begin{cases} 1, & \text{if } y_2 \in Z_{2j}, \\ 0, & \text{otherwise}, \end{cases} \quad j = 0, 1, 2, \quad (3.2.53)$$

we may express $I_{10}(y_1)$ as

$$I_{10}(y_1) = \int_{Z_{20}} P_1(k-1)\Lambda_{11}p(y_2|H_1) + P_2(k-1)\Lambda_{12}p(y_2|H_2)$$
$$+ \int_{Z_{21}} P_1\Lambda_{11}p(y_2|H_1) + P_2(k-1)\Lambda_{12}p(y_2|H_2)$$
$$+ \int_{Z_{22}} P_1(k-1)\Lambda_{11}p(y_2|H_1) + P_2\Lambda_{12}p(y_2|H_2).$$

(3.2.54)

Similarly,

$$I_{11}(y_1) = \int_{Z_{20}} P_0 p(y_2|H_0) + P_2(k-1)\Lambda_{12}p(y_2|H_2)$$
$$+ \int_{Z_{21}} P_0(k-1)p(y_2|H_0) + P_2(k-1)\Lambda_{12}p(y_2|H_2)$$
$$+ \int_{Z_{22}} P_0(k-1)p(y_2|H_0) + P_2\Lambda_{12}p(y_2|H_2),$$

(3.2.55)

and

$$I_{12}(y_1) = \int_{Z_{20}} P_0 p(y_2|H_0) + P_1(k-1)\Lambda_{11} p(y_2|H_1)$$
$$+ \int_{Z_{21}} P_0(k-1) p(y_2|H_0) + P_1 \Lambda_{11} p(y_2|H_1)$$
$$+ \int_{Z_{22}} P_0(k-1) p(y_2|H_0) + P_1(k-1)\Lambda_{11} p(y_2|H_1).$$

(3.2.56)

Substitution of these in the decision rule (3.2.49a) and some simplification yields

$$\int_{Z_{20}} P_1(k-1)\Lambda_{11} p(y_2|H_1) + \int_{Z_{21}} P_1 \Lambda_{11} p(y_2|H_1) + \int_{Z_{22}} P_1(k-1)\Lambda_{11} p(y_2|H_1)$$

$$\begin{array}{c} H_1 \text{ or } H_2 \\ > \\ < \\ H_0 \text{ or } H_2 \end{array} \int_{Z_{20}} P_0 p(y_2|H_0) + \int_{Z_{21}} P_0(k-1) p(y_2|H_0) + \int_{Z_{22}} P_0(k-1) p(y_2|H_0).$$

(3.2.57)

Letting $Z_{21} = Z - Z_{20} - Z_{22}$ on the left-hand side, $Z_{20} = Z - Z_{21} - Z_{22}$ on the right-hand side, and rearranging, the decision rule becomes

$$\Lambda_{11} \begin{array}{c} H_1 \text{ or } H_2 \\ > \\ < \\ H_0 \text{ or } H_2 \end{array} \frac{P_0}{P_1} \frac{(k-1)-(k-2)\int_{Z_{20}} p(y_2|H_0)}{(k-1)-(k-2)\int_{Z_{21}} p(y_2|H_1)} \triangleq t_{11}. \quad (3.2.58)$$

Similarly, we can obtain the other two decision rules:

$$\Lambda_{12} \begin{array}{c} H_0 \text{ or } H_2 \\ > \\ < \\ H_0 \text{ or } H_1 \end{array} \Lambda_{11} \frac{P_1}{P_2} \frac{(k-1)-(k-2)\int_{Z_{21}} p(y_2|H_1)}{(k-1)-(k-2)\int_{Z_{22}} p(y_2|H_2)}. \quad (3.2.59)$$

and

$$\Lambda_{12} \underset{H_0 \text{ or } H_1}{\overset{H_1 \text{ or } H_2}{\gtrless}} \frac{P_0}{P_2} \frac{(k-1)-(k-2)\int_{Z_{20}} p(y_2|H_0)}{(k-1)-(k-2)\int_{Z_{22}} p(y_2|H_2)} \triangleq t_{12}. \quad (3.2.60)$$

The inequalities (3.2.58) to (3.2.60) specify the decision rule at detector 1. A similar set of inequalities can be obtained for detector 2. The decision rules at the two detectors are coupled. The decision regions at detector 1 can be represented in the $(\Lambda_{11}, \Lambda_{12})$ plane. The decision rules at detector 2 are similar in nature and are represented in the $(\Lambda_{21}, \Lambda_{22})$ plane. Equations (3.2.58) to (3.2.60) and the corresponding set for detector 2 need to be solved simultaneously to determine the decision regions at the two detectors. For the special cost assignment $k = 2$, decision rules at the two detectors decouple, and the resulting decision regions are the same as obtained for the single-sensor three-hypothesis detection problem [Van68]. The decision rules at detector 1 for the $k = 2$ case are given by

$$\Lambda_{11} \underset{H_0 \text{ or } H_2}{\overset{H_1 \text{ or } H_2}{\gtrless}} \frac{P_0}{P_1}, \quad (3.2.61a)$$

$$\Lambda_{12} \underset{H_0 \text{ or } H_1}{\overset{H_0 \text{ or } H_2}{\gtrless}} \frac{P_1}{P_2} \Lambda_{11}, \quad (3.2.61b)$$

and

$$\Lambda_{12} \underset{H_0 \text{ or } H_1}{\overset{H_1 \text{ or } H_2}{\gtrless}} \frac{P_0}{P_2}. \quad (3.2.61c)$$

Decision rules for cost assignments other than of the form given in (3.2.51) can be determined in a similar manner.

3.3 Design of Fusion Rules

In the distributed detection system without fusion considered in Section 3.2, decisions were made locally and were not transmitted to a fusion center for decision combining. The effect of fusion could be incorporated by special cost assignments, but the design of fusion rules was not investigated. In this section, we consider the companion problem, namely, combining or fusion of local decisions.

Consider again the binary hypothesis testing problem with hypothesis H_0 and H_1 and their associated prior probabilities P_0 and P_1. There are N local detectors, each making a local decision u_i, $i = 1, ..., N$, where

$$u_i = \begin{cases} 0, & \text{if detector } i \text{ decides } H_0, \\ 1, & \text{otherwise.} \end{cases} \quad (3.3.1)$$

As shown in Figure 3.5, local decisions form the input to the fusion center which combines them to yield the global decision u_0,

$$u_0 = \begin{cases} 0, & \text{if } H_0 \text{ is decided,} \\ 1, & \text{otherwise.} \end{cases} \quad (3.3.2)$$

The problem is to determine the fusion rule to combine local decisions based on some optimization criterion.

The fusion rule is essentially a logical function with N binary inputs and one binary output. In general, there are 2^{2^N} fusion rules when there are N binary inputs to the fusion center. For example, there are 16 possible fusion rules for combining two binary local decisions as listed in Table 3.1. Note that commonly used logical functions AND and OR are 2 of the 16 logical functions. The fusion rule f_2 represents the AND rule in which $u_0 = 1$ only when both of the local decisions are 1, i.e.,

$$u_0 = \begin{cases} 1, & \text{if } u_1 = 1 \text{ and } u_2 = 1, \\ 0, & \text{otherwise.} \end{cases} \quad (3.3.3)$$

60   3. Distributed Bayesian Detection: Parallel Fusion Network

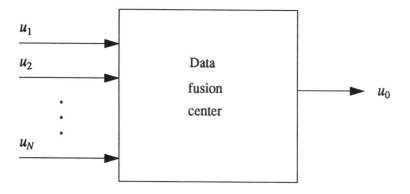

Figure 3.5. Fusion center.

Table 3.1. Possible Fusion Rules for Two Binary Decisions

Input	Output u_0															
$u_1\ u_2$	f_1	f_2	f_3	f_4	f_5	f_6	f_7	f_8	f_9	f_{10}	f_{11}	f_{12}	f_{13}	f_{14}	f_{15}	f_{16}
0 0	0	0	0	0	0	0	0	0	1	1	1	1	1	1	1	1
0 1	0	0	0	0	1	1	1	1	0	0	0	0	1	1	1	1
1 0	0	0	1	1	0	0	1	1	0	0	1	1	0	0	1	1
1 1	0	1	0	1	0	1	0	1	0	1	0	1	0	1	0	1

3.3 Design of Fusion Rules 61

The OR rule is given by f_8 in which $u_0 = 1$ as long as at least one of the local decisions is 1, i.e.,

$$u_0 = \begin{cases} 0, & \text{if } u_1 = 0 \text{ and } u_2 = 0, \\ 1, & \text{otherwise}. \end{cases} \quad (3.3.4)$$

Several of the sixteen functions may not be suitable as fusion rules. For example, the all zero function f_1 and the all one function f_{16} totally disregard the two inputs and, therefore, may not be preferred in many situations. Similarly, some other functions that may be unsuitable totally disregard one of the inputs, e.g., the function f_4 totally disregards u_2. Later in this section, we shall develop the monotonicity property of optimum fusion rules that will reduce the size of the set of permissible fusion rules from 2^{2^N} to a much smaller number.

An ad hoc approach to the design of fusion rules would be to arbitrarily select a fusion rule from the set of commonly used logical functions, such as the AND, the MAJORITY, and the OR functions. An arbitrary choice of the fusion rule may provide satisfactory results in many instances, but optimality is not guaranteed. Next, we consider the design of optimum fusion rules in a Bayesian framework. Each input to the fusion center is a binary random variable characterized by the associated probabilities of false alarm and miss. Let P_{Fi}, P_{Di} and P_{Mi} denote the probabilities of false alarm, detection and miss of detector i respectively, i.e.,

$$P_{Fi} = P(u_i = 1 | H_0),$$

$$P_{Di} = P(u_i = 1 | H_1),$$

and

$$P_{Mi} = P(u_i = 0 | H_1).$$

The probabilities of false alarm, detection, and miss of the overall system, i.e., at the fusion center, denoted by P_F, P_D, and P_M, are given by

$$P_F = P(u_0 = 1 | H_0),$$

$$P_D = P(u_0 = 1|H_1),$$
and
$$P_M = P(u_0 = 0|H_1).$$

The objective is to determine the fusion rule or decision rule at the fusion center that minimizes the average cost. The problem can be viewed as a two hypothesis detection problem with individual detector decisions being the observations. The optimum fusion rule is given by the following likelihood ratio test:

$$\frac{P(u_1, u_2, ..., u_N | H_1)}{P(u_1, u_2, ..., u_N | H_0)} \begin{array}{c} u_0 = 1 \\ > \\ < \\ u_0 = 0 \end{array} \frac{P_0(C_{10} - C_{00})}{P_1(C_{01} - C_{11})} \triangleq \eta, \qquad (3.3.5)$$

where C_{ij} denotes the cost of global decision being H_i, when H_j is present. Due to the independence of local decisions, the left-hand side (LHS) of (3.3.5) can be written as

$$\frac{P(u_1, u_2, ..., u_N | H_1)}{P(u_1, u_2, ..., u_N | H_0)} = \prod_{i=1}^{N} \frac{P(u_i | H_1)}{P(u_i | H_0)}$$

$$= \prod_{S_1} \frac{P(u_i = 1 | H_1)}{P(u_i = 1 | H_0)} \prod_{S_0} \frac{P(u_i = 0 | H_1)}{P(u_i = 0 | H_0)}$$

$$= \prod_{S_1} \frac{1 - P_{Mi}}{P_{Fi}} \prod_{S_0} \frac{P_{Mi}}{1 - P_{Fi}}, \qquad (3.3.6)$$

where S_j is the set of all those local decisions that are equal to j, $j = 0, 1$. Substituting (3.3.6) in (3.3.5) and taking the logarithm of both sides,

$$\sum_{S_1} \log \frac{1 - P_{Mi}}{P_{Fi}} + \sum_{S_0} \log \frac{P_{Mi}}{1 - P_{Fi}} \begin{array}{c} u_0 = 1 \\ > \\ < \\ u_0 = 0 \end{array} \log \eta, \qquad (3.3.7)$$

or

$$\sum_{i=1}^{N}\left[u_i \log \frac{1-P_{Mi}}{P_{Fi}} + (1-u_i) \log \frac{P_{Mi}}{1-P_{Fi}}\right] \begin{matrix} u_0=1 \\ > \\ < \\ u_0=0 \end{matrix} \log \eta. \qquad (3.3.8)$$

This rule can also be expressed as

$$\sum_{i=1}^{N}\left[\log \frac{(1-P_{Mi})(1-P_{Fi})}{P_{Mi}P_{Fi}}\right] u_i \begin{matrix} u_0=1 \\ > \\ < \\ u_0=0 \end{matrix} \log \left[\eta \prod_{i=1}^{N}\left(\frac{1-P_{Fi}}{P_{Mi}}\right)\right]. \qquad (3.3.9)$$

Thus, the optimum fusion rule can be implemented by forming a weighted sum of the incoming local decisions and, then, comparing it with a threshold. The weights and the threshold are determined by the reliability of the decisions, i.e., by the probabilities of miss and false alarm of the local detectors. The threshold also depends on the prior probabilities and the costs.

Remark 3.3.1

The optimum fusion rule is monotonic under the assumption that $P_{Di} \geq P_{Fi}$, $i = 1, \ldots, N$. This is a reasonable assumption for meaningful detection problems. A monotonic fusion rule is defined as follows. Let $S_1(k)$ be the set of k local decisions that are equal to one and $S_0(N-k)$ be the set of remaining $(N-k)$ local decisions that are equal to zero. Given $S_1(k)$ and $S_0(N-k)$, let the global decision be $u_0 = 1$. Let $S_1(k')$, $k' > k$, be another set of k' local decisions that are equal to one, so that $S_1(k')$ contains $S_1(k)$, and let $S_0(N-k')$ be the corresponding set of local decisions equal to zero. For a fusion rule to be monotonic, the global decision is $u_0 = 1$ for all possible sets $S_1(k')$ satisfying the above property.

Given $S_1(k)$ and $S_0(N-k)$, the optimum fusion rule can be written as

$$\prod_{S_1(k)} \frac{1-P_{Mi}}{P_{Fi}} \prod_{S_0(N-k)} \frac{P_{Mi}}{1-P_{Fi}} \begin{matrix} u_0=1 \\ > \\ < \\ u_0=0 \end{matrix} \eta. \qquad (3.3.10)$$

To have $u_0 = 1$, the LHS of (3.3.10) must be greater than η. Because $P_{Di} \geq P_{Fi}$,

$$\frac{1-P_{Mi}}{P_{Fi}} \geq 1, \quad i = 1, \ldots, N,$$

and

$$\frac{P_{Mi}}{1-P_{Fi}} \leq 1, \quad i = 1, \ldots, N.$$

Consequently, for any $S_1(k')$ and $S_0(N - k')$, $k' > k$, such that $S_1(k')$ contains $S_1(k)$,

$$\prod_{S_1(k')} \frac{1-P_{Mi}}{P_{Fi}} \prod_{S_0(N-k')} \frac{P_{Mi}}{1-P_{Fi}} \geq \prod_{S_1(k)} \frac{1-P_{Mi}}{P_{Fi}} \prod_{S_0(N-k)} \frac{P_{Mi}}{1-P_{Fi}} \geq \eta$$

(3.3.11)

and the corresponding global decision is $u_0 = 1$. Thus, the optimum fusion rule is monotonic. This concept is similar to the concept of monotonicity of switching functions. This property of fusion rules drastically reduces the number of fusion rules to be considered during the design process. This reduction is shown in Table 3.2. Certain other considerations or constraints often result in further reductions of the number of fusion rules to be considered.

Remark 3.3.2

For a fusion center with any fixed threshold η and any fixed monotonic fusion rule, the probability of detection P_D is an increasing function of P_{Di}, $i = 1, \ldots, N$. This has been proven in [TVB89] by taking advantage of the fact that the optimum fusion rule is a monotonic and increasing Boolean function. Consequently, the optimum fusion rule can be denoted by a form that is a sum of products only of u_i, $i = 1, \ldots, N$, and no complements $\bar{u}_i$ appear. Because u_i, $i = 1, \ldots, N$, are independent of each other, P_D can be expressed in terms of P_{Di}, and, by taking the partial derivative of P_D with respect to P_{Di}, $i = 1, \ldots, N$, the result follows.

Table 3.2. Number of Monotonic Fusion Rules

Number of detectors N	Number of possible fusion rules 2^{2^N}	Number of monotonic fusion rules
2	16	6
3	256	20
4	65,536	168

Example 3.3

Consider the fusion of two local decisions. Let the probabilities of false alarm and miss be given by $P_{Fi} = P_{Mi} = 0.1$, $i = 1, 2$. Let us consider the minimum probability of error cost function, i.e., $C_{00} = C_{11} = 0$, and $C_{01} = C_{10} = 1$.

There are 16 possible fusion rules as given in Table 3.1. Out of these, only f_1, f_2, f_4, f_6, f_8, and f_{16} are monotonic and need to be considered further. In this example, the two decisions have been assumed to have identical statistics and, therefore, u_0 should be identical for both $u_1 = 0$, $u_2 = 1$ and $u_1 = 1$, $u_2 = 0$ cases. Thus, f_4 and f_6 are not suitable fusion rules. Finally, the optimum fusion rule is determined from (3.3.9) for different values of the prior probability P_0 and is given in Table 3.3. In this example, f_1 and f_{16} turn out to be optimal for certain values of P_0.

Fusion Rule for the Soft Decision Case

In this section thus far, we have considered the design of fusion rules for the case when the local detectors could make only hard decisions, i.e., u_i could take only two values 0 or 1 corresponding to the two hypotheses H_0 and H_1. Next, we generalize the design of fusion rules for the binary hypothesis testing problem to include the case in which the local detectors are allowed to make multilevel or soft decisions. Assume that the observation space at each local detector is partitioned into J mutually exclusive regions so that, if the observation at detector i lies in the partition j, we set $u_i = j$, $j = 0, ..., J - 1$. These soft decisions are transmitted to the fusion center which combines them to yield the global decision u_0 as defined earlier in (3.3.2). Let us define the following probabilities:

Table 3.3. Optimum Fusion Rules for Example 3.3

Range of values for P_0	Optimum fusion rule
0 – 0.012	f_{16} (all ones)
0.012 – 0.5	f_8 (OR)
0.5 – 0.988	f_2 (AND)
0.988 – 1	f_1 (all zeros)

$$\alpha_{ji} = P(u_i = j|H_0), \qquad (3.3.12)$$

and

$$\beta_{ji} = P(u_i = j|H_1). \qquad (3.3.13)$$

The fusion rule that minimizes the average cost is still given by (3.3.5) but now the LHS can be expressed as

$$\frac{P(u_1, u_2, ..., u_N|H_1)}{P(u_1, u_2, ..., u_N|H_0)} = \prod_{i=1}^{N} \frac{P(u_i|H_1)}{P(u_i|H_0)}$$

$$= \prod_{j=0}^{J-1} \prod_{S_j} \frac{P(u_i = j|H_1)}{P(u_i = j|H_0)}$$

$$= \prod_{j=0}^{J-1} \prod_{S_j} \frac{\beta_{ji}}{\alpha_{ji}}, \qquad (3.3.14)$$

where S_j is the set of all those local decisions u_i that are equal to j, $j = 0, 1, ..., J - 1$. Substituting (3.3.14) in (3.3.5) and taking the logarithm of both sides, we obtain the minimum average cost fusion rule as

$$\sum_{j=0}^{J-1} \sum_{S_j} \log\left(\frac{\beta_{ji}}{\alpha_{ji}}\right) \underset{u_0=0}{\overset{u_0=1}{\gtrless}} \log \eta. \qquad (3.3.15)$$

This fusion rule is an obvious generalization of the fusion rule for the hard decision case. Once again, the fusion rule depends on the reliabilities of the local detectors, i.e., their probabilities of errors.

Fusion Rule with Direct Observations

The fusion center considered thus far accepts the set of hard or soft decisions from the local detectors. It can not observe the phenomenon directly. Next, we consider the case where the fusion center also receives direct observations regarding the phenomenon. Let y_0 denote the direct observation at the fusion center. The inputs to the fusion center are a continuous random variable y_0 and the discrete random vector u. The optimum fusion rule in this case is given by the following likelihood ratio test:

$$\frac{p(y_0|H_1)}{p(y_0|H_0)} \frac{P(u_1,u_2,...,u_N|H_1)}{P(u_1,u_2,...,u_N|H_0)} \underset{u_0=0}{\overset{u_0=1}{\gtrless}} \eta. \qquad (3.3.16)$$

Using (3.3.14) and denoting the likelihood ratio test for the observation y_0 as $\Lambda(y_0)$,

$$\Lambda(y_0) \prod_{j=0}^{J-1} \prod_{S_j} \frac{\beta_{ji}}{\alpha_{ji}} \underset{u_0=0}{\overset{u_0=1}{\gtrless}} \eta. \qquad (3.3.17)$$

This decision rule can also be expressed as

$$\Lambda(y_0) \underset{u_0=0}{\overset{u_0=1}{\gtrless}} \eta \prod_{j=0}^{J-1} \prod_{S_j} \frac{\alpha_{ji}}{\beta_{ji}}, \qquad (3.3.18)$$

where the threshold on the RHS is a function of the decision vector u and can take J^N possible values. Taking the logarithm of both sides, we may express the fusion rule in another form, a generalized version of (3.3.15),

$$\log \Lambda(y_0) + \sum_{j=0}^{J-1} \sum_{S_j} \log \left(\frac{\beta_{ji}}{\alpha_{ji}}\right) \begin{matrix} u_0 = 1 \\ > \\ < \\ u_0 = 0 \end{matrix} \log \eta. \quad (3.3.19)$$

This fusion rule minimizes the average cost in the case where the fusion center can observe the phenomenon directly, and it also receives soft decisions from the local detectors.

Fusion Rule for Asynchronous Decisions

Thus far, it has been assumed that local decisions arrive at the fusion center in a synchronous manner, i.e., they are based on a common clock and arrive at some predetermined instants. As discussed in [ChK 94], this assumption may not be valid in some situations, and local decisions may arrive at the fusion center in an asynchronous manner. Here, we develop the fusion rule for combining asynchronous hard decisions. In the rest of the book, however, only synchronous systems will be considered.

It is assumed that the distributed detection system operates over a given observation interval $[0,\tau]$ during which the hypothesis present does not change. Based on their observations, local detectors make (a multiple number of) decisions about the hypothesis present at asynchronous instants during $[0,\tau]$. As these decisions are made, they are transmitted to the fusion center. The fusion center combines incoming local decisions as they arrive and finally yields the global decision at the end of the observation interval $[0,\tau]$. Incoming decisions are assumed to be conditionally statistically independent (both spatial and temporal independence) under hypothesis $H_j, j = 0, 1$. We assume that the number of decisions that each detector makes over the observation interval is Poisson distributed. This implies that the total number of local decisions that arrive at the fusion center over $[0,\tau]$ are also Poisson distributed. Let λ_{0i} and λ_{1i}, $i = 1, ..., N$, denote the average number of decisions per unit time made by the ith detector under H_0 and H_1 respectively. Then, the conditional probability that the ith detector makes $k_{\tau i}$ decisions during the observation interval $[0,\tau]$ is given by

$P(k_{\tau i}$ decisions during $[0,\tau]|H_j)$

$$= \frac{e^{-\lambda_{ji}\tau}(\lambda_{ji}\tau)^{k_{\tau i}}}{k_{\tau i}!}, \quad i = 1, ..., N; \; j = 0, 1; \; k_{\tau i} = 0, 1... \; . \qquad (3.3.20)$$

The set of local decisions transmitted by DM i during $[0,\tau]$ are denoted by $\boldsymbol{u}_i = \{u_{i1}, u_{i2}, ..., u_{ik_{\tau i}}\}$. Note that it is possible for this set to be null for one or more DM i, i.e., one or more DM i may not transmit any decisions at all during $[0,\tau]$. The probabilities of false alarm and detection corresponding to DM i, P_{Fi} and P_{Di}, are assumed to be stationary, i.e., they are not functions of time. The global decision is made at the end of the observation interval $[0,\tau]$ based on all the local decisions received during this period. The optimum fusion rule for this system is given by

$$\frac{P(\boldsymbol{u}_1, ..., \boldsymbol{u}_N|H_1)}{P(\boldsymbol{u}_1, ..., \boldsymbol{u}_N|H_0)} \begin{matrix} u_0^\tau = 1 \\ > \\ < \\ u_0^\tau = 0 \end{matrix} \eta, \qquad (3.3.21)$$

where u_0^τ denotes the global decision at the end of the observation interval, i.e., at time τ. Due to conditional independence, (3.3.21) becomes

$$\prod_{i=1}^{N} \frac{P(\boldsymbol{u}_i|H_1)}{P(\boldsymbol{u}_i|H_0)} \begin{matrix} u_0^\tau = 1 \\ > \\ < \\ u_0^\tau = 0 \end{matrix} \eta.$$

As indicated earlier, the number of elements in each $\boldsymbol{u}_i$ is a Poisson random variable. Using Bayes rule,

$$\prod_{i=1}^{N} \frac{P(\boldsymbol{u}_i|k_{\tau i}, H_1)}{P(\boldsymbol{u}_i|k_{\tau i}, H_0)} \frac{P(k_{\tau i}|H_1)}{P(k_{\tau i}|H_0)} \begin{matrix} u_0^\tau = 1 \\ > \\ < \\ u_0^\tau = 0 \end{matrix} \eta. \qquad (3.3.22)$$

Substituting for $P(k_{\tau i}|H_j)$, $j = 0, 1$, and taking the natural logarithm of both sides,

$$\sum_{i=1}^{N} \log \frac{P(u_i|k_{\tau i}, H_1)}{P(u_i|k_{\tau i}, H_0)} \underset{u_0^\tau = 0}{\overset{u_0^\tau = 1}{\underset{<}{>}}} \log \eta + \sum_{i=1}^{N} (\lambda_{1i} - \lambda_{0i})\tau$$
$$+ \sum_{i=1}^{N} k_{\tau i} \log \frac{\lambda_{0i}}{\lambda_{1i}}. \quad (3.3.23)$$

The left-hand side can be expressed as

$$\sum_{i=1}^{N} \sum_{k=1}^{k_{\tau i}} \left[u_k \log \frac{P_{Di}}{P_{Fi}} + (1 - u_k) \log \frac{1 - P_{Di}}{1 - P_{Fi}} \right],$$

and the optimum fusion rule becomes

$$\sum_{i=1}^{N} \left[k_{\tau i} \log \frac{1 - P_{Di}}{1 - P_{Fi}} + \sum_{k=1}^{k_{\tau i}} \log \frac{P_{Di}(1 - P_{Fi})}{P_{Fi}(1 - P_{Di})} \right] \underset{u_0^\tau = 0}{\overset{u_0^\tau = 1}{\underset{<}{>}}}$$
$$\log \eta + \sum_{i=1}^{N} (\lambda_{1i} - \lambda_{0i})\tau + \sum_{i=1}^{N} k_{\tau i} \log \frac{\lambda_{01}}{\lambda_{1i}}. \quad (3.3.24)$$

In the special case of synchronous decisions, this fusion rule reduces to the one derived earlier. Also, note that this fusion rule is a function of the observation interval. For example, if the observation interval is increased, the global decision may change even if no new local decisions are received.

Fusion of Correlated Decisions

Next, we consider the fusion of correlated local decisions. For simplicity, attention is limited to the case of hard decisions. In this case again, the optimum fusion rule is given by (3.3.5), but the statistic on the LHS includes correlation of local decisions.

Drakopoulos and Lee [DrL91] have shown that the fusion rule in this case is given by

$$\frac{\sum_{I \subseteq S_0} (-1)^{|I|} E_1 \left(\prod_{i \in S_1 \cup I} u_i \right)}{\sum_{I \subseteq S_0} (-1)^{|I|} E_0 \left(\prod_{i \in S_1 \cup I} u_i \right)} \begin{array}{c} u_0 = 1 \\ > \\ < \\ u_0 = 0 \end{array} \eta, \qquad (3.3.25)$$

where S_0, S_1 are as defined earlier, $I \subseteq \{1, 2, ..., N\}$ with $I \neq \phi$, $|I|$ is the cardinality of the set I, and E_j, $j = 0, 1$, denotes expectation under hypothesis H_j. This fusion rule reduces to (3.3.5) when the local decisions are independent.

Kam et al. [KZG92] employed another approach, namely, the Bahadur–Lazarfeld expansion of probability density functions [DuH73, pp. 111–113] to show that the optimum fusion rule for correlated local decisions can be expressed as the following generalization of (3.3.9):

$$\sum_{i=1}^{N} \left[\log \frac{(1-P_{Mi})(1-P_{Fi})}{P_{Mi} P_{Fi}} \right] u_i + \sum_{i=1}^{N} \log \left(\frac{P_{Mi}}{1-P_{Fi}} \right)$$

$$+ \log \frac{1 + \sum_{i<j} K_{ij}^1 z_i^1 z_j^1 + \sum_{i<j<k} K_{ijk}^1 z_i^1 z_j^1 z_k^1 + ... + K_{12\_n}^1 z_1^1 z_2^1 ... z_n^1}{1 + \sum_{i<j} K_{ij}^0 z_i^0 z_j^0 + \sum_{i<j<k} K_{ijk}^0 z_i^0 z_j^0 z_k^0 + ... + K_{12\_n}^0 z_1^0 z_2^0 ... z_n^0}$$

$$\begin{array}{c} u_0 = 1 \\ > \\ < \\ u_0 = 0 \end{array} \log \eta, \qquad (3.3.26)$$

where

$$z_i^h = \frac{u_i - P(u_i = 1|H_h)}{\sqrt{P(u_i = 1|H_h)[1 - P(u_i = 1|H_h)]}}, \quad h = 0, 1,$$

$$K_{ij}^h = \sum_u z_i^h z_j^h P(u|H_h),$$

$$K_{ijk}^h = \sum_u z_i^h z_j^h z_k^h P(u|H_h),$$

$$\vdots$$

$$K_{12\_n}^h = \sum_u z_1^h z_2^h .. z_n^h P(u|H_h).$$

In many practical situations, conditional correlation coefficients beyond a certain order can be assumed to be zero. In that case, computation of the LHS of the optimum fusion rule (3.3.26) becomes less burdensome. When all the conditional correlation coefficients are zero, the optimum fusion rule (3.3.26) reduces to (3.3.9).

3.4 Detection with Parallel Fusion Network

The two components of the detection problem for the parallel fusion network topology, namely, detection by distributed detectors based on a coupled cost assignment and fusion of incoming local decisions have been considered in Sections 3.2 and 3.3, respectively. Here, we consider the binary hypothesis testing problem for the overall system shown in Figure 3.6. All the local detectors DM i, $i = 1, ..., N$, observe the same phenomenon. The observations of the local detectors are denoted by y_i, $i = 1, ..., N$, and their joint conditional density $p(y_1, ..., y_N|H_j)$, $j = 0, 1$, is assumed known. No communication among local detectors is assumed. Based on its own observation y_i, each local detector makes a local decision u_i, $i = 1, ..., N$. Local decisions are transmitted over bandlimited channels to the data fusion center, where they are combined to yield the global inference. As in (3.2.1), each local decision u_i may take the value 0 or 1, depending on whether the detector DM i decides H_0 or H_1. Note that the only case considered at this point is the one in

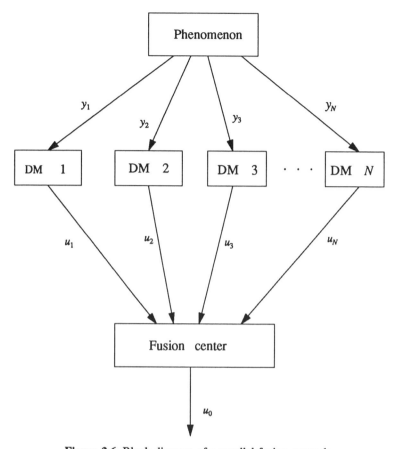

Figure 3.6. Block diagram of a parallel fusion network.

which hard decisions are transmitted from the local detectors. The case of soft decision systems, where the local observation space is partitioned into more than two regions and this information is sent to the fusion center, is considered later in this chapter. Other data compression schemes at local detectors may also be implemented, but these alternate systems are not treated in this book. The fusion center yields the global decision u_0 based on the received decision vector containing the local decisions, i.e., $u^T = (u_1, ..., u_N)$. The global decision u_0 is assumed to depend only on the local decision vector u and not on the observations y_i at the individual detectors. The probabilities of false alarm, miss, and detection corresponding to detector DM i are denoted by P_{Fi}, P_{Mi}, and P_{Di}, respectively, and are as defined in Section 3.3. Similarly, the overall probabilities of false alarm, miss, and detection are denoted by P_F, P_M

and P_D, respectively, and are again defined in Section 3.3. These probabilities characterize system performance and will be useful in system optimization.

The objective in the Bayesian hypothesis testing problem for the parallel fusion network is to obtain the set of decision rules $\Gamma=\{\gamma_0, \gamma_1, ..., \gamma_N\}$ that minimizes the average cost of the overall system operation $\Re(\Gamma)$. The fusion rule is denoted by γ_0 and the decision rules at the local detectors are denoted by γ_i, $i = 1, ..., N$. These decision rules are mappings from the observation space to the decision space, i.e.,

$$u_i = \gamma_i(y_i), \quad i = 1..., N, \tag{3.4.1}$$

and

$$u_0 = \gamma_0(u_1, ..., u_N). \tag{3.4.2}$$

For notational simplicity, we do not explicitly show the dependence of $\Re$ on Γ unless it is required. The Bayes risk function that we wish to minimize can be written as

$$\Re = \sum_{i=0}^{1} \sum_{j=0}^{1} C_{ij} P_j P(\text{Decide } H_i | H_j \text{ is present}), \tag{3.4.3}$$

where C_{ij} is the cost of global decision being H_i when H_j is present and P_j is the a priori probability of hypothesis H_j, $i, j = 0, 1$. Grouping some terms, the risk function can be expressed as

$$\Re = C_F P_F - C_D P_D + C, \tag{3.4.4}$$

where

$$C_F = P_0(C_{10} - C_{00}),$$

$$C_D = (1 - P_0)(C_{01} - C_{11}),$$

and

$$C = C_{01}(1 - P_0) + C_{00} P_0.$$

3.4 Detection with Parallel Fusion Network

We make the usual assumption that making a wrong decision is more costly than making a correct decision, i.e., $C_{10} > C_{00}$, and $C_{01} > C_{11}$. This implies that $C_F > 0$ and $C_D > 0$. The overall probabilities of false alarm and detection can be expressed as

$$P_F = \sum_{u} P(u_0 = 1 | u) P(u | H_0), \qquad (3.4.5)$$

and

$$P_D = \sum_{u} P(u_0 = 1 | u) P(u | H_1), \qquad (3.4.6)$$

where $\sum_{u}$ indicates summation over all possible values of the decision vector u. Substituting (3.4.5) and (3.4.6) in (3.4.4),

$$\Re = C + C_F \sum_{u} P(u_0 = 1 | u) P(u | H_0) - C_D \sum_{u} P(u_0 = 1 | u) P(u | H_1). \qquad (3.4.7)$$

The system is designed next by obtaining decision rules that attempt to minimize $\Re$.

Derivation of Decision Rules

For system optimization, a person-by-person optimization (PBPO) methodology for this team decision problem is adopted. The decentralized detection system of Figure 3.6 is viewed as a team consisting of two members. The fusion center is one team member, whereas the aggregation of the local detectors is the other team member. The second team member can be further viewed as a team consisting of local detectors as team members within their aggregate team. While optimizing one team member, it is assumed that the other team members have already been designed and remain fixed. System design equations resulting from this PBPO procedure represent necessary but not, in general, sufficient conditions to determine the globally optimum solution. This set of equations is solved simultaneously to obtain the desired PBPO solution.

First, we obtain the local decision rules that minimize $\Re$. While deriving the decision rule at a specific local detector, say DM k, it will be assumed that the fusion center and all the other local detectors have already been designed and remain fixed. We expand $\Re$ explicitly in terms of the kth local decision, u_k.

$$\Re = C + \sum_{u^k} \left\{ P(u_0 = 1 | u^{k1}) \left[C_F P(u^{k1} | H_0) - C_D P(u^{k1} | H_1) \right] \right. $$
$$\left. + P(u_0 = 1 | u^{k0}) \left[C_F P(u^{k0} | H_0) - C_D P(u^{k0} | H_1) \right] \right\}, \quad (3.4.8)$$

where

$$u^k = (u_1, \ldots, u_{k-1}, u_{k+1}, \ldots, u_N)^T,$$

and

$$u^{kj} = (u_1, \ldots, u_{k-1}, u_k = j, u_{k+1}, \ldots, u_N)^T, \quad j = 0, 1.$$

By noting that

$$P(u^{k0} | H_j) = P(u^k | H_j) - P(u^{k1} | H_j),$$

we may express $\Re$ as

$$\Re = C + \sum_{u^k} \left\{ P(u_0 = 1 | u^{k1}) \left[C_F P(u^{k1} | H_0) - C_D P(u^{k1} | H_1) \right] \right.$$
$$+ P(u_0 = 1 | u^{k0}) \left[C_F P(u^k | H_0) - C_D P(u^k | H_1) \right.$$
$$\left. \left. - C_F P(u^{k1} | H_0) + C_D P(u^{k1} | H_1) \right] \right\}$$
$$= C_k + \sum_{u^k} A(u^k) \left[C_F P(u^{k1} | H_0) - C_D P(u^{k1} | H_1) \right], \quad (3.4.9)$$

where

$$C_k = C + \sum_{u^k} P(u_0 = 1 | u^{k0}) \left[C_F P(u^k | H_0) - C_D P(u^k | H_1) \right],$$

and

$$A(u^k) = P(u_0 = 1 | u^{k1}) - P(u_0 = 1 | u^{k0}).$$

We may express

$$P(u | H_j) = \int_Y P(u | Y) p(Y | H_j) dY,$$

where $Y = (y_1, ..., y_N)^T$ and $\int_Y \cdot$ represents a multifold integral over all components of Y. Since the decision of each detector depends only on its own observation,

$$P(u | Y) = \prod_{i=1}^{N} P(u_i | y_i), \qquad (3.4.10)$$

and

$$P(u^{ki} | Y) = P(u_k = i | y_k) P(u^k | Y^k), \quad i = 0, 1, \quad (3.4.11)$$

we may write $P(u^{ki} | H_j)$ as

$$\begin{aligned} P(u^{ki} | H_j) &= \int_Y P(u^{ki} | Y) p(Y | H_j) dY \\ &= \int_Y P(u_k = i | y_k) P(u^k | Y^k) p(Y | H_j) dY, \quad (3.4.12) \end{aligned}$$

where

$$Y^k = (y_1, ..., y_{k-1}, y_{k+1}, ..., y_N)^T.$$

Thus,

$$\Re = C_k + \sum_{u^k} A(u^k)[C_F \int_Y P(u_k=1|y_k)P(u^k|Y^k)p(Y|H_0)dY$$

$$- C_D \int_Y P(u_k=1|y_k)P(u^k|Y^k)p(Y|H_1)dY]$$

$$= C_k + \int_{y_k} dy_k P(u_k=1|y_k)$$

$$\times \left\{ \sum_{u^k} \int_{Y^k} dY^k A(u^k)P(u^k|Y^k)[C_F p(Y|H_0) - C_D p(Y|H_1)] \right\}.$$

(3.4.13)

At this point, we assume that all the detectors other than DM k remain fixed and obtain the decision rule at DM k so as to minimize $\Re$. As far as DM k is concerned, C_k is a constant, and we obtain the following decision rule

$$P(u_k=1|y_k) = \begin{cases} 1, & \text{if } D(k) \leq 0, \\ 0, & \text{otherwise}, \end{cases} \quad (3.4.14)$$

where

$$D(k) = \sum_{u^k} \int_{Y^k} dY^k\, A(u^k)\, P(u^k|Y^k)\, [C_F\, p(Y|H_0) - C_D\, p(Y|H_1)].$$

Noting that

$$p(Y|H_j) = p(Y^k|y_k, H_j) p(y_k|H_j), \quad (3.4.15)$$

the decision rule at DM k can be expressed in an alternate form as

$$p(y_k|H_1) \sum_{u^k} \int_{Y^k} A(u^k)\, C_D P(u^k|Y^k)\, p(Y^k|y_k, H_1)\, dY^k$$

$$\sum_{u_k=0}^{u_k=1} p(y_k|H_0) \sum_{u^k} \int_{Y^k} A(u^k) C_F P(u^k|Y^k) \, p(Y^k|y_k, H_0) \, dY^k.$$

(3.4.16)

Next, we obtain the fusion rule that minimizes $\Re$. Following the PBPO methodology, we assume that all the local detectors have been designed and remain fixed. Based on their observations, local detectors yield the decision vector u. The conditional distributions $P(u|H_j)$, $j = 0, 1$, are assumed to be known. Because the elements of u are binary valued, there are 2^N possible values of u. Let u^* be one out of 2^N possible values of u. We express $\Re$ from (3.4.7) in terms of u^* as follows:

$$\Re = P(u_0 = 1|u^*) [C_F P(u^*|H_0) - C_D P(u^*|H_1)] + K(u^*), \tag{3.4.17}$$

where

$$K(u^*) = C + \sum_{\substack{u \\ u \neq u^*}} P(u_0 = 1|u)[C_F P(u|H_0) - C_D P(u|H_1)].$$

For a given u^*, $K(u^*)$ is fixed. Therefore, $\Re$ is minimized if we employ the following fusion rule:

$$P(u_0 = 1|u^*) = \begin{cases} 1, & \text{if } [C_F P(u^*|H_0) - C_D P(u^*|H_1)] \leq 0, \\ 0, & \text{otherwise}. \end{cases} \tag{3.4.18}$$

Using the fact that $C_F > 0$ and $C_D > 0$, we may also express the fusion rule as

$$C_D P(u^*|H_1) - C_F P(u^*|H_0) \quad \underset{P(u_0=1|u^*)=0}{\overset{P(u_0=1|u^*)=1}{\gtrless}} \quad 0, \tag{3.4.19}$$

or

$$\frac{P(u^*|H_1)}{P(u^*|H_0)} \underset{u_0=0}{\overset{u_0=1}{\underset{<}{>}}} \frac{C_F}{C_D}. \qquad (3.4.20)$$

Thus, we have 2^N equations of the form (3.4.20) and N equations of the form (3.4.16). A simultaneous solution of these $(N + 2^N)$ nonlinear coupled equations yields the PBPO solution to the binary decentralized Bayesian hypothesis testing problem. The hypothesis testing problem with M hypotheses can be solved similarly. Results are not provided here and can be found in [Hob86]. The number of equations to be solved in this case becomes $(M - 1)N + M^N$. This number grows rapidly with the number of detectors N, and the computational effort becomes prohibitive. Computation becomes easier for some special cases. These cases for the binary hypothesis testing problem are discussed next.

Special Cases

Conditionally Independent Local Observations

The local decision rules specified in (3.4.16) are not, in general, threshold tests. If, however, the observations at the local detectors are assumed to be conditionally independent, the local decision rules reduce to threshold tests. Under the conditional independence assumption,

$$p(Y|H_j) = \prod_{i=1}^{N} p(y_i|H_j), \quad j = 0, 1. \qquad (3.4.21)$$

Therefore, we may express the local decision rule from (3.4.16) as

$$\frac{p(y_k|H_1)}{p(y_k|H_0)} \underset{u_k=0}{\overset{u_k=1}{\underset{<}{>}}} \frac{\sum_{u^k} C_F A(u^k) \prod_{i=1,i\neq k}^{N} \int_{Y^i} P(u_i|y_i) p(y_i|y_k, H_0) dy_i}{\sum_{u^k} C_D A(u^k) \prod_{i=1,i\neq k}^{N} \int_{Y^i} P(u_i|y_i) p(y_i|y_k, H_1) dy_i},$$

$$(3.4.22)$$

which reduces to

$$\frac{p(y_k|H_1)}{p(y_k|H_0)} \overset{u_k=1}{\underset{u_k=0}{\gtrless}} \frac{\sum_{u^k} C_F A(u^k) \prod_{i=1,i\neq k}^{N} P(u_i|H_0)}{\sum_{u^k} C_D A(u^k) \prod_{i=1,i\neq k}^{N} P(u_i|H_1)}. \qquad (3.4.23)$$

We observe that the right-hand side of (3.4.23) is a constant, and, thus, the local decision rules are threshold tests. The fusion rule of (3.4.20) for conditionally independent observations can be expressed as

$$\prod_{i=1}^{N} \frac{P(u_i|H_1)}{P(u_i|H_0)} \overset{u_0=1}{\underset{u_0=0}{\gtrless}} \frac{C_F}{C_D}. \qquad (3.4.24)$$

The number of coupled nonlinear equations to be solved still remains the same, but the computational difficulty is reduced because the local decision rules reduce to threshold tests. Also, the fusion rule given in (3.4.24) can be implemented by forming a weighted sum and comparing it to a threshold as given in (3.3.8).

Example 3.4

In this example, we obtain explicit expressions for the thresholds and the fusion rule for the parallel fusion network with two local detectors and conditionally independent observations. From (3.4.23), the threshold at DM 1 is given by

$$t_1 = \frac{\sum_{u_2} C_F A(u_2) P(u_2|H_0)}{\sum_{u_2} C_D A(u_2) P(u_2|H_1)}$$

$$= \frac{C_F}{C_D} \frac{P_{F2}(P_{100} - P_{110} + P_{111} - P_{101}) + (P_{110} - P_{100})}{P_{M2}(P_{110} - P_{100} + P_{101} - P_{111}) + (P_{111} - P_{101})},$$

where

$$P_{ijk} = P(u_0 = i | u_1 = j, u_2 = k).$$

Similarly, the threshold at DM 2 is given by

$$t_2 = \frac{C_F}{C_D} \frac{P_{F1}(P_{100} - P_{101} + P_{111} - P_{110}) + (P_{101} - P_{100})}{P_{M1}(P_{101} - P_{100} + P_{110} - P_{111}) + (P_{111} - P_{110})}.$$

In this problem, the decision vector u can assume four values. The fusion rule is specified by (3.4.24) and is given by the following four relationships:

$$\frac{P_{M1}}{1-P_{F1}} \frac{P_{M2}}{1-P_{F2}} \underset{u_0=0}{\overset{u_0=1}{\gtrless}} \frac{C_F}{C_D},$$

$$\frac{P_{M1}}{1-P_{F1}} \frac{1-P_{M2}}{P_{F2}} \underset{u_0=0}{\overset{u_0=1}{\gtrless}} \frac{C_F}{C_D},$$

$$\frac{1-P_{M1}}{P_{F1}} \frac{P_{M2}}{1-P_{F2}} \underset{u_0=0}{\overset{u_0=1}{\gtrless}} \frac{C_F}{C_D},$$

and

$$\frac{1-P_{M1}}{P_{F1}} \frac{1-P_{M2}}{P_{F2}} \underset{u_0=0}{\overset{u_0=1}{\gtrless}} \frac{C_F}{C_D}.$$

The above six nonlinear equations are coupled, and their simultaneous solution yields the result.

To obtain more specific results, assume that the conditional densities at the two detectors are Gaussian. Also, assume that the minimum probability of error cost assignment is employed. Under H_0, the conditional densities at the two detectors are assumed to be identical with mean zero and variance one. Under H_1, the mean at DM j, $j = 1$, 2, is given by m_j and the variance at both detectors is unity. The set of six nonlinear equations is solved to obtain the thresholds and the fusion rule for three different sets of values $m_1 = 1$, $m_2 = 1$; $m_1 = 1$, $m_2 = 1.5$; and $m_1 = 1$, $m_2 = 2$. The resulting optimum fusion rules are shown in Table 3.4. They depend upon the value of P_0 and are shown using the notation from Table 3.1. For unequal values of m_1 and m_2, the optimum fusion rule is to accept the decision of DM 2 for certain values of P_0. Thresholds at the two detectors and the resulting probability of error for two sets of values of m_1 and m_2 are shown in Figures 3.7 – 3.10. Note the discontinuities in the threshold values. These are due to the use of different fusion rules for different ranges of P_0. At these discontinuities, either fusion rule can be employed. Fusion rules corresponding to different ranges of P_0 are indicated in Figures 3.8 and 3.10. In the regions where the fusion rule is to accept the decision of DM 2, thresholds at DM 1 are not shown because they do not affect system performance. In the case, $m_1 = 1$, $m_2 = 1$, the thresholds at the two detectors are found to be identical. Probability of error curves are shown for four fusion rules f_2, f_4, f_6 and f_8. The system probability of error curve is obtained by forming the lower envelope of the probability of error curves corresponding to different fusion rules.

Receiver operating characteristics (ROC) corresponding to different fusion rules for the two sets of values of m_1 and m_2 are shown in Figures 3.11 and 3.12. These curves again show the superiority of different fusion rules in different regions. The ROC of the overall optimum system or the team ROC consists of ROC segments corresponding to best fusion rules in different regions, i.e., it consists of the upper envelopes of different ROC curves. For example, in Figure 3.11, the team ROC consists of the ROC segment corresponding to the OR rule for smaller values of P_F and the ROC segment corresponding to the AND rule for larger values of P_F. Note that the ROC corresponding to individual fusion rules are concave but the team ROC is not concave due to a switch in the fusion rule employed. This is seen more clearly in Figure 3.13 by zooming in on the appropriate portion of the ROC for the case $m_1 = .3$, $m_2 = 3$.

84   3. Distributed Bayesian Detection: Parallel Fusion Network

Table 3.4. Optimum Fusion Rules for Example 3.4

m_1, m_2	Range of values for P_0	Optimum fusion rule
$m_1 = 1$ $m_2 = 1$	0 − 0.5 0.5 − 1.0	f_8 (OR) f_2 (AND)
$m_1 = 1$ $m_2 = 1.5$	0 − 0.185 0.185 − 0.5 0.5 − 0.815 0.815 − 1.0	f_6 (same as DM 2) f_8 (OR) f_2 (AND) f_6 (same as DM 2)
$m_1 = 1$ $m_2 = 2$	0 − 0.38 0.38 − 0.5 0.5 − 0.62 0.62 − 1.0	f_6 (same as DM 2) f_8 (OR) f_2 (AND) f_6 (same as DM 2)

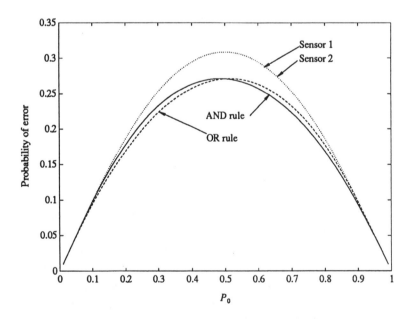

Figure 3.7. Probability of error for Example 3.4, $m_1 = m_2 = 1$.

3.4 Detection with Parallel Fusion Network   85

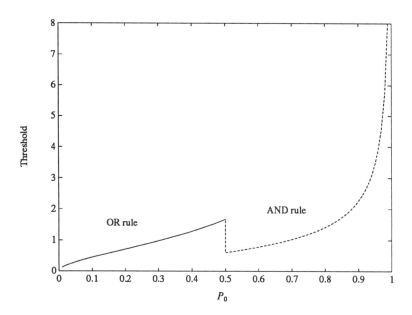

Figure 3.8. Threshold values for Example 3.4, $m_1 = m_2 = 1$.

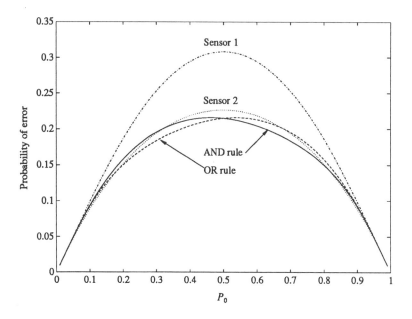

Figure 3.9. Probability of error for Example 3.4, $m_1 = 1$, $m_2 = 1.5$.

86   3. Distributed Bayesian Detection: Parallel Fusion Network

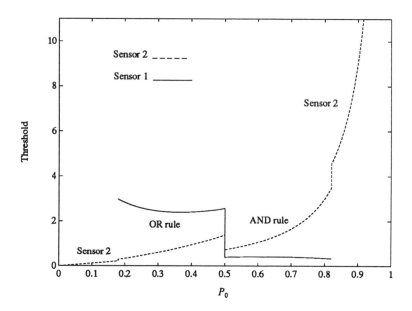

Figure 3.10. Threshold values for Example 3.4, $m_1 = 1$, $m_2 = 1.5$.

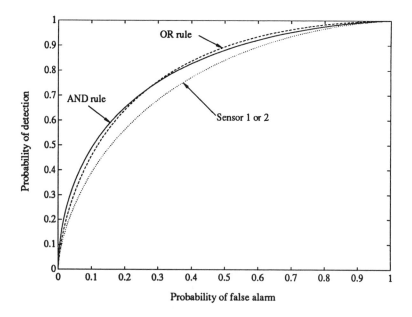

Figure 3.11. Receiver operating characteristics for Example 3.4, $m_1 = m_2 = 1$.

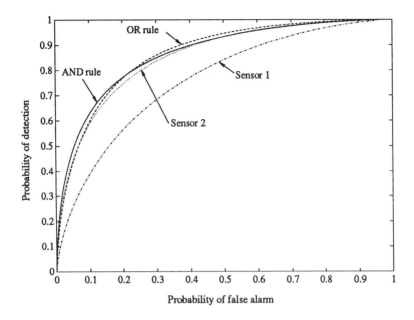

Figure 3.12. Receiver operating characteristics for Example 3.4, $m_1=1$, $m_2=1.5$.

Remark 3.4.1

In this remark, we discuss concavity of the team ROC. This discussion is limited to continuous ROC. By allowing randomization, this discussion can be suitably modified to include discontinuous ROC. Recall that a continuous ROC passes through points $P_F = P_D = 0$ and $P_F = P_D = 1$. Also, the slope of the tangent of an ROC at any operating point is equal to the value of the threshold required to operate at that point. All possible values of the threshold (determined by the prior probabilities and cost assignment) are attained at some point of the ROC. In the Bayesian team detection problem, all possible values of the threshold (and slope) are attained on the concave segments of the team ROC. Consider the common tangent of the two individual ROCs as shown in Figure 3.13 and let O and A be points as shown. Slope values greater than the slope of the common tangent are attained on the team ROC segment corresponding to the AND fusion rule, i.e., on the segment linking points $P_F = P_D = 0$ and A. When the two slope values are equal, either of the two points O or A can be used. Slope values less than the slope of the common tangent are attained on the team ROC segment linking points O and $P_F = P_D = 1$. Thus, the threshold deter-

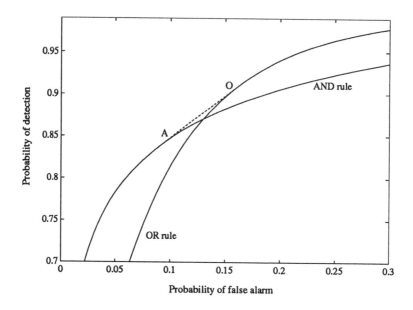

Figure 3.13. Illustration of nonconcavity of team ROC, $m_1 = .3$, $m_2 = 3$.

mined by any cost assignment and values of prior probabilities yields an operating point that lies on the concave segments of the team ROC. This naturally means that certain values of P_F and P_D (corresponding to the region of the common tangent) are not achieved. This is not a requirement in the Bayesian formulation anyway. However, if it is desired to attain certain specified values of P_F and P_D (as in Neyman–Pearson detection to be discussed in Chapter 5) that are not achievable by the above Bayesian formulation, randomization may need to be performed. A side benefit of randomization is that it makes the team ROC concave. In the randomized decision rule, strategy at point A is used with probability δ and, with probability $(1 - \delta)$, strategy at point O is used. By an appropriate choice of δ, any operating point on the common tangent can be attained, i.e., any desirable value of P_F or P_D that was previously unattainable can now be attained. If (α_a, d_a) and (α_o, d_o) represent the probabilities of false alarm and detection at points A and O, respectively, the randomized decision rule yields

$$P_F = \delta \alpha_a + (1 - \delta) \alpha_o, \quad P_F \in (\alpha_a, \alpha_o),$$

and

$$P_D = \delta d_a + (1 - \delta)d_o, \ P_D \in (d_a, d_o).$$

Team ROC for the randomized decision rule consists of the previous two concave segments and the common tangent. It is concave. Note that at both O and A the slope of the common tangent is the same, but they employ different thresholds. Readers will also note that the randomization described above is dependent randomization [Tsi93a, Pap 90, WiW 92] and requires synchronization among decision makers. Individual DMs employ the same value of δ which needs to be known prior to decision making. This additional communication, which may be carried out off-line, introduces dependence in randomization. To a certain extent, this randomization procedure works against the desired objective of decentralization but is required to achieve some specified goals, e.g., to attain a specified value of P_F. It should be pointed out that there is another randomization procedure, namely, independent randomization. In this procedure, each DM determines its own randomization parameter δ independently of others. This procedure does not necessarily make the team ROC concave.

Identical Local Detectors

Another special case of interest occurs when the observations at the local detectors are conditionally independent as well as identically distributed. For a symmetric cost assignment, the initial inclination is to use identical decision rules at the local detectors. In Example 3.4, the thresholds at the local detectors turned out to be identical for the case $m_1 = m_2 = 1$. However, identical decision rules are not optimum, in general. This situation was observed in the distributed detection systems without fusion considered in Examples 3.1 and 3.2. When the cost of making two errors k was large, the optimum action was to employ nonidentical decision rules. It is reasonable to use different decision rules to avoid making two errors. Examples showing the nonoptimality of identical decision rules for distributed detection systems with fusion are presented in Tsitsiklis [Tsi86], Willett and Warren [WiW91], and Cherikh and Kantor [ChK92] among others. In most such cases, discrete or nearly discrete density functions have been employed. We present the example from Tsitsiklis [Tsi86] for illustration.

Example 3.5

Consider a parallel fusion network consisting of two local detectors and a fusion center. Let H_0 and H_1 denote two equally likely hypotheses under consideration. The local observations y_1 and y_2 are assumed to be conditionally independent under the two hypotheses. The observations are discrete and take values in the set $\{1,2,3\}$. The conditional distributions are identical and are given by

$$p(y=1|H_0) = 4/5, \quad p(y=2|H_0) = 1/5, \quad p(y=3|H_0) = 0,$$

$$p(y=1|H_1) = 1/3, \quad p(y=2|H_1) = 1/3, \quad p(y=3|H_1) = 1/3.$$

The objective is to determine the decision rules at the local detectors and the fusion center so as to minimize the overall probability of error. Because the local observations are conditionally independent, each detector performs a threshold test. There are only two candidate threshold tests that can be implemented at the local detectors.

Decision rule A:

$$u_i = \begin{cases} 0, & \text{if } y_i = 1, \\ 1, & \text{otherwise}. \end{cases} \quad (3.4.25)$$

Decision rule B:

$$u_i = \begin{cases} 0, & \text{if } y_i = 1, 2, \\ 1, & \text{otherwise}. \end{cases} \quad (3.4.26)$$

Three possibilities need to be considered. There are two possibilities in which both detectors employ identical decision rules, i.e., either both detectors employ decision rule A or both detectors employ decision rule B. The third possibility involves the use of nonidentical decision rules at the two detectors, i.e., one detector, say DM 1, uses decision rule A and the other uses decision rule B. There are six monotonic fusion rules to be considered here to determine the optimum set of decision rules. Fusion rules f_1 and f_{16} that completely disregard u_1 and u_2 are not considered any further in this example. An exhaustive enumeration approach is used to determine the set of optimum decision rules. If decision rule A is used at a detector,

$$P_{Fi} = P(u_i = 1|H_0) = 1/5,$$

$$P_{Mi} = P(u_i = 0|H_1) = 1/3.$$

When decision rule B is used at a detector,

$$P_{Fi} = P(u_i = 1|H_0) = 0,$$

$$P_{Mi} = P(u_i = 0|H_1) = 2/3.$$

When fusion rules f_4 or f_6 are employed, the global decision is set equal to the incoming decision from DM 1 or DM 2. In this case, the global probabilities of false alarm and miss are determined by the decision rule in use (A or B) at the detector whose decision is accepted. The global probabilities of false alarm and miss for the AND and OR fusion rules are given as follows.

AND fusion rule:

$$P_F = P_{F1}P_{F2}, \qquad (3.4.27a)$$

$$P_M = P_{M1} + P_{M2} - P_{M1}P_{M2}. \qquad (3.4.27b)$$

OR fusion rule:

$$P_F = P_{F1} + P_{F2} - P_{F1}P_{F2}, \qquad (3.4.28a)$$

$$P_M = P_{M1}P_{M2}. \qquad (3.4.28b)$$

The overall probability of error for the case of equally likely hypotheses is given by

$$P(e) = \tfrac{1}{2}(P_F + P_M).$$

The probability of error for all the cases is listed in Table 3.5. The minimum occurs when one detector uses decision rule A, the other one

Table 3.5. Probability of Error Enumeration for Example 3.5

	Local Decision Rules Used at DM 1 and DM 2		
Fusion rule	AA	AB	BB
f_2 (AND)	67/225	7/18	4/9
f_4 (Accept u_1)	4/15	4/15	1/3
f_6 (Accept u_2)	4/15	1/3	1/3
f_8 (OR)	53/225	19/90	2/9

uses decision rule B, and the fusion rule is the OR rule. Thus, the optimum solution is to employ nonidentical local decision rules in spite of identical observation statistics.

The result of Example 3.5 is rather surprising. In spite of the fact that the local observations are identically distributed and the costs are symmetric, the optimum solution is to use nonidentical decision rules at the local detectors. This counterexample makes the computation of optimum local decision rules intractable especially when the number of local detectors becomes large. This is due to the fact that a search over all nonidentical local decision rules is required. If local decision rules were constrained to be identical, it would result in a drastic reduction in computational effort. Willett and Warren [WiW91] investigated this problem and derived a necessary condition under which local decision rules are identical when observations at all the sensors are independent and identically distributed. Their result provides an intuitive explanation for the counterexamples and shows that, in these cases, local likelihood ratios exhibit discrete or point-mass like behavior. Irving and Tsitsiklis [IrT94] have considered a more specific binary hypothesis testing problem. They consider a two sensor system in which hard decisions are transmitted to the fusion center. Noise corrupting the observations at the two sensors is assumed to be independent with an identical Gaussian distribution. It is analytically shown that no optimality is lost if the thresholds for the LRTs at the sensors are constrained to be identical. This result depends upon the special structure of the problem. Similar analytical results for more general problems are not available at this time.

3.4 Detection with Parallel Fusion Network 93

Tsitsiklis [Tsi88] has further considered the issue of identical decision rules for the binary hypothesis testing problem from an asymptotic point of view. From the well-known results of Chernoff [Che52], it is easy to see that, for any reasonable set of decision rules, the probability of error for the parallel fusion network goes to zero exponentially as the number of local detectors increases to infinity. To discriminate between different sets of decision rules as N → ∞, Tsitsiklis [Tsi88] has considered the exponent of the error probability given by

$$r_N(\gamma^N) = \frac{\log \mathfrak{R}_N(\gamma^N)}{N}, \quad (3.4.29)$$

where $r_N(\cdot)$ is the error exponent for the N-detector problem, $\gamma^N = (\gamma_1, ..., \gamma_N)$ is the set of local decision rules and $\mathfrak{R}_N(\cdot)$ is the probability of error at the fusion center for the N-detector problem. It was shown that, for the binary hypothesis testing problem, as $N \to \infty$, the error exponents of the optimal solution and the solution obtained by constraining the local decision rules to be identical are equal, i.e., the solution with identical local decision rule constraint is asymptotically optimum. Furthermore, it was shown that for an M-hypothesis problem, the local decision rules can be restricted to a set consisting of $M(M-1)/2$ distinct decision rules to obtain an asymptotically optimum solution. For the binary hypothesis testing problem, Chen and Papamarcou [ChP93] have investigated the exact asymptotics of the minimum error probabilities achieved by the optimum parallel fusion network and the system obtained by invoking the identical local decision rule constraint. They showed analytically that the local decision rules of the optimum system are almost identical and are, therefore, marginally different from decision rules obtained under the identical decision rule constraint. Numerical experience [Tsi93a] also shows that the restriction of identical decision rules leads to little or no loss of performance. These results provide a justification for using the identical decision rule constraint, in general. As pointed out earlier, this constraint reduces the amount of computational effort drastically. For a detailed discussion of asymptotic results and proofs, the reader is referred to [Tsi88, ChP93].

We further consider the case in which the local decision rules are restricted to be identical. From the monotonicity of optimum fusion rules, it can easily be seen that the fusion rule in this case reduces to a K-out-of-N fusion rule, i.e., the global decision $u_0 = 1$ if K or more local decisions are equal to 1. This simplifies the computation even further. Next, we obtain the optimum value of K that minimizes the risk $\Re$ for the K-out-of-N fusion rule.

Because the local detectors have been assumed to be identical, the probabilities of false alarm and detection at each of the local detectors are the same. We let $P_{Di} = p_D$ and $P_{Fi} = p_F$, $i = 1, ..., N$. Using this notation, the overall probabilities of detection and false alarm can be expressed as

$$P_D = \sum_{i=K}^{N} \binom{N}{i} (p_D)^i (1-p_D)^{N-i}, \qquad (3.4.30)$$

and

$$P_F = \sum_{i=K}^{N} \binom{N}{i} (p_F)^i (1-p_F)^{N-i}, \qquad (3.4.31)$$

where

$$\binom{N}{i} = \frac{N!}{i!(N-i)!}.$$

Substituting for P_D and P_F in (3.4.4), we may express $\Re$ as a function of K as

$$\Re(K) = C + \sum_{i=K}^{N} \binom{N}{i} \left[C_F(p_F)^i (1-p_F)^{N-i} - C_D(p_D)^i (1-p_D)^{N-i} \right].$$

$$(3.4.32)$$

When K is increased by 1, the increment in $\Re$ can be written as

$$\Re(K+1) - \Re(K) = \binom{N}{K} \left[C_D(p_D)^K (1-p_D)^{N-K} - C_F(p_F)^K (1-p_F)^{N-K} \right].$$

$$(3.4.33)$$

Under the assumption $p_D \geq p_F$, it can easily be seen that $\Re(\cdot)$ has a single minimum. From (3.4.33), $\Re(K+1) \geq \Re(K)$ if

$$\left(\frac{p_D}{p_F}\right)^K \left(\frac{1-p_D}{1-p_F}\right)^{N-K} \geq \frac{C_F}{C_D} \qquad (3.4.34)$$

and $\Re(K+1) < \Re(K)$, otherwise. Taking the logarithm of both sides of (3.4.34), $\Re(K+1) \geq \Re(K)$ if

$$K \log \frac{p_D(1-p_F)}{p_F(1-p_D)} + N \log \frac{1-p_D}{1-p_F} \geq \log \frac{C_F}{C_D} \qquad (3.4.35)$$

and $\Re(K+1) < \Re(K)$, otherwise. Rearranging terms, we can write $\Re(K+1) \geq \Re(K)$ if

$$K \geq \frac{\log\left\{\left(\frac{C_F}{C_D}\right)\left[(1-p_F)/(1-p_D)\right]^N\right\}}{\log\left\{p_D(1-p_F)/p_F(1-p_D)\right\}} \triangleq K^* \qquad (3.4.36)$$

and $\Re(K+1) < \Re(K)$, otherwise.

Thus, the optimum value of K, K_{opt}, for the K-out-of-N fusion rule is given by

$$K_{opt} = \begin{cases} \lceil K^* \rceil, & \text{if } K^* \geq 0, \\ 0, & \text{otherwise,} \end{cases} \qquad (3.4.37)$$

where $\lceil \cdot \rceil$ denotes the standard ceiling function. It should be pointed out that K-out-of-N fusion rules can be implemented in terms of logic functions. Commonly used fusion rules, such as AND, OR, or MAJORITY are N-out-of-N, 1-out-of-N, and $\lceil (N+1)/2 \rceil$-out-of-N rules, respectively. These rules are optimum only for certain ranges of parameter values, but they are used widely due to their simplicity of implementation.

Example 3.6

Consider a parallel fusion network consisting of three local detectors with conditionally independent and identically distributed observations. Assume that the minimum probability of error cost assignment is employed.

First, we let the conditional densities at the local detectors be Gaussian with unit variance. Under H_0 and H_1, the mean at all the detectors is assumed to be zero and one, respectively. In this identical distribution case, the local thresholds are constrained to be equal. The fusion rule is of the K-out-of-N form, and only the OR, the MAJORITY, and the AND fusion rules need to be considered. For each fusion rule, the value of the local threshold as a function of P_0 is determined by solving a nonlinear equation in t_i similar to the ones shown in Example 3.4 for the two detector case. The probability of error of the optimum system as a function of P_0 is shown in Figure 3.14. As indicated, for $0 \leq P_0 \leq .09$, the OR fusion rule is the best; for $.91 \leq P_0 \leq 1$, the AND fusion rule yields the minimum value of the probability of error. For all other values of P_0, the MAJORITY rule performs the best. The corresponding values of the threshold are shown in Figure 3.15. Once again, discontinuities in the threshold values indicate switching of fusion

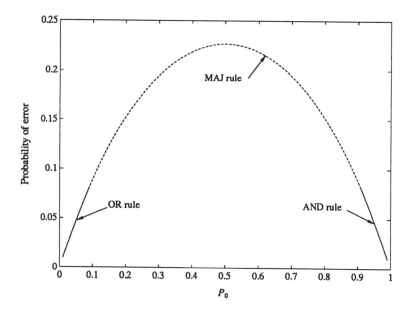

Figure 3.14. Probability of error for Example 3.6, Gaussian case.

3.4 Detection with Parallel Fusion Network 97

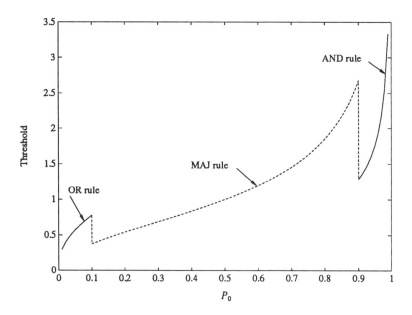

Figure 3.15. Threshold values for Example 3.6, Gaussian case.

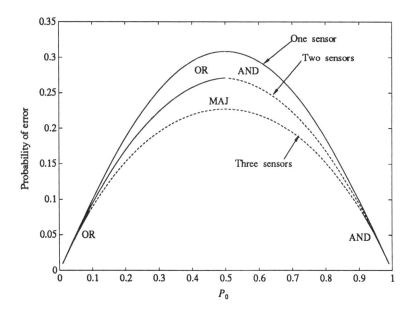

Figure 3.16. Performance as a function of number of detectors.

98 3. Distributed Bayesian Detection: Parallel Fusion Network

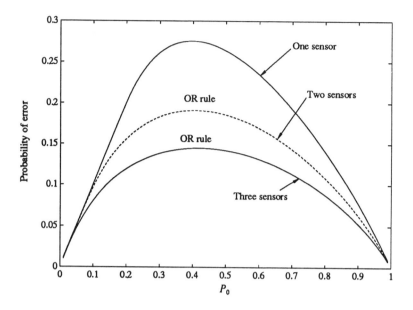

Figure 3.17. Probability of error for Example 3.6, Rayleigh case.

rules. In Figure 3.16, we show the performance enhancement, as detectors are added, by plotting the best achievable probabilities of error as a function of P_0 for parallel fusion networks consisting of one, two, and three local detectors. Observation statistics at each detector are assumed to be identical.

Next, we let the conditional densities at the local detectors be Rayleigh given by

$$p(y_i|H_j) = \frac{y_i}{\sigma_j^2} \exp\left(-\frac{y_i^2}{2\sigma_j^2}\right), \quad i = 1, 2, 3; \quad j = 0, 1; \quad y_i \geq 0.$$

Let $\sigma_0 = 1$ and $\sigma_1 = 2$. Once again, local thresholds are constrained to be identical. In this case, the OR fusion rule is found to be the best for all values of P_0. In Figure 3.17, we plot the probability of error of the optimum system as a function of P_0 and also observe the effect of adding identical detectors to the system.

Example 3.7

This example treats the case of discrete observations at the local detectors. Consider the parallel fusion network with two local detectors and conditionally independent, identically distributed observations. Let the conditional densities at the two detectors be Poisson given by

$$p(n|H_i) = \frac{(m_i)^n}{n!} \exp(-m_i), \quad n = 0, 1, \ldots; \quad i = 0, 1,$$

where m_i represents the mean under hypothesis H_i. The likelihood ratio at each detector is given by

$$\Lambda(n) = \left(\frac{m_1}{m_0}\right)^n \exp\left[-(m_1 - m_0)\right].$$

We constrain the thresholds at the two detectors to be identical and determine the ROC for the OR and the AND fusion rules. Assume that the minimum probability of error cost assignment is employed and let $m_0 = 6$, and $m_1 = 9$. The likelihood ratio test at each detector is of the form

$$\Lambda(n) \underset{u_i = 0}{\overset{u_i = 1}{\gtrless}} t,$$

which reduces to

$$n \underset{u_i = 0}{\overset{u_i = 1}{\gtrless}} \frac{\log(t) + m_1 - m_0}{\log(m_1) - \log(m_0)} \triangleq \eta.$$

Let

$$\eta^* = \lceil \eta \rceil.$$

In this case,

$$P_{Fi} = p_F = 1 - \exp(-m_0) \sum_{n=0}^{\eta^*-1} \frac{(m_0)^n}{n!},$$

and

$$P_{Di} = p_D = 1 - \exp(-m_1) \sum_{n=0}^{\eta^*-1} \frac{(m_1)^n}{n!}.$$

The global probabilities of false alarm and detection for the two fusion rules can be obtained as follows:

AND fusion rule:

$$P_F = p_F^2,$$

$$P_D = p_D^2.$$

OR fusion rule:

$$P_F = 1 - (1 - p_F)^2,$$

$$P_D = 1 - (1 - p_D)^2.$$

Resulting ROCs are shown in Figure 3.18. Note that, in this case, ROCs consist of discrete points, and only certain values of the pair (P_F, P_D) are achieved. Of course, randomized decision rules can be employed at the local detectors. By using the thresholds corresponding to two adjacent discrete points along with an appropriate randomization parameter, any point on the straight lines joining the discrete points can be achieved. In Figure 3.18, we have shown ROCs that include the straight line segments achievable by randomization. Also, note that the ROCs corresponding to the two fusion rules intersect each other. The team ROC again consists of the upper envelopes of the two individual ROCs, and it can be made concave by randomization of fusion rules in the manner discussed in Remark 3.4.1.

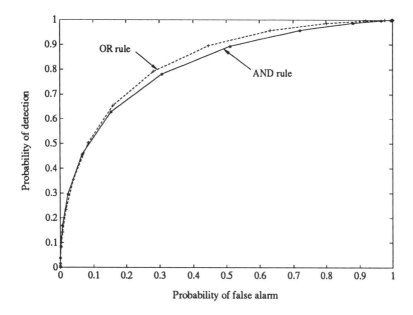

Figure 3.18. Receiver operating characteristics for Example 3.7.

Computational Algorithms

The PBPO solution to the binary decentralized Bayesian hypothesis testing problem using a parallel fusion network is obtained by solving the $(N + 2^N)$ necessary conditions. Another computational approach is to obtain the solution by an exhaustive search over all possible decision strategies. For each fusion rule, one may obtain the set of local thresholds and the associated detection cost for the system. The fusion rule and the corresponding set of local thresholds that yield the minimum cost constitute the solution. This approach can be employed for relatively small networks, but it becomes impractical as N becomes large. In fact, by assuming that the observation space at the local detectors is discrete (continuous observations at the local detectors can be discretized for an exhaustive search), Tsitsiklis [Tsi93a] has shown that the computational complexity of determining the optimal strategy is exponential in N, the number of sensors. If, however, the local detectors are identical and are restricted to employ identical decision rules, the complexity is polynomial in N. In any case, an exhaustive search approach for networks of a reasonable size is impractical. A more efficient approach is to pick a decision strategy to start with and employ an algorithm, such as the Gauss–Seidel cyclic coordinate descent algorithm, to obtain the solution in an iterative manner [Tan90, TPK91a]. This algorithm is discussed next.

The algorithm is initialized by picking an initial decision strategy $\Gamma^{(0)} = \{\gamma_0^{(0)}, \gamma_1^{(0)}, ..., \gamma_N^{(0)}\}$, i.e., by picking a set of local thresholds and a fusion rule. The superscript zero indicates initialized decision strategies. Other values of superscripts indicate the decision strategy determined during that iteration, e.g., $\gamma_i^{(j)}$ indicates decision strategy γ_i determined during iteration j. During the first iteration, first, the decision rule at the fusion center $\gamma_0^{(1)}$ is obtained so as to minimize $\Re(\gamma_0, \gamma_1^{(0)}, ..., \gamma_N^{(0)})$ while the other decision strategies $\gamma_1^{(0)},...,\gamma_N^{(0)}$ are held fixed. Next, $\gamma_1^{(1)}$ is obtained so as to minimize $\Re(\gamma_0^{(1)}, \gamma_1, \gamma_2^{(0)}, ..., \gamma_N^{(0)})$. Similarly, all the decision rules $\gamma_i^{(1)}$, $i = 2, ..., N$, are obtained by minimizing $\Re(\gamma_0^{(1)}, \gamma_1^{(1)}, ..., \gamma_{i-1}^{(1)}, \gamma_i, \gamma_{i+1}^{(0)}, ..., \gamma_N^{(0)})$. Once the first iteration is complete, the next iterations are run identically. The iterative algorithm terminates when the difference in the values of the Bayes risk attained at the end of two successive iterations is less than a prespecified tolerance, ε. Specifically, the algorithm is given below.

Step 1: Initialize the algorithm by selecting the local thresholds $t_i^{(0)}$, $i = 1, ..., N$, and the fusion rule $\gamma_0^{(0)}$. Compute the probabilities $P^{(0)}(u_i|H_j)$, all i, j. Set the iteration index $n = 0$.

Step 2: (Main iteration)

Step 2.1: Determine the fusion rule $\gamma_0^{(n)}$ using (3.4.24) based on $P^{(n)}(u_i|H_j)$, all i, j.

Step 2.2: For each DM i, $i = 1, ..., N$, compute threshold $t_i^{(n)}$ using (3.4.23) with updated values of $A(u^k)$ and $P(u_i|H_j)$ based on threshold values $t_1^{(n)}, t_2^{(n)}, ..., t_{i-1}^{(n)}, t_{i+1}^{(n-1)}, ..., t_N^{(n-1)}$.

Step 3: Compute the Bayes risk $\Re(\Gamma^{(n)})$. If $|\Re(\Gamma^{(n)}) - \Re(\Gamma^{(n-1)})| < \varepsilon$, STOP.
Else, set $n = n + 1$, and return to Step 2.

Under the assumption that the probability density function of the likelihood ratios of the local detectors is continuous and nonzero over the domain of definition, $\Re(\Gamma^{(n)})$ has been shown to be nonincreasing [Tan90]. Therefore, if $\Re(\Gamma^{(n)})$ does not change appreciably during two successive iterations, the algorithm has converged to a person-by-person optimal solution satisfying the necessary conditions of optimality. At this point, the directional derivative of $\Re(\Gamma^{(n)})$ along any feasible direction is non-negative, and the point is either a local minimum or a saddle point. Even though the cost does not change during two successive iterations, there is no guarantee that the decision rules also converge. The result depends upon the choice of the initial strategies, and it may converge to different solutions for different initial choices. In spite of some of its shortcomings, practical experience shows that this algorithm is quite successful and converges in a few iterations. A number of examples are presented in [TPK91a] where a number of issues have been addressed, e.g., the effect of different starting points, variations in costs, number and quality of local detectors, etc. These examples reaffirm the important observation regarding person-by-person optimal solutions that, whenever one team member changes its decision strategy, other team members should be informed, so that they can adapt their decision strategies to improve the overall system performance.

Tang [Tan90] has further considered the issue of computational algorithms for the design of parallel fusion networks. He has shown that the problem can be formulated as a minimization problem in terms of the N false alarm probabilities P_{Fi}, $i = 1, ..., N$, of the local detectors, or, equivalently, in terms of the local thresholds. This nonlinear programming problem with rectangular (box) constraints can be stated as

$$\min_{F} \Re(\Gamma) = \min_{F} \ C_F P_F - C_D P_D + C \qquad (3.4.38)$$

subject to

$$0 \leq P_{Fi} \leq 1 \quad \text{for } 1 \leq i \leq N,$$

where

$$F = (P_{F1}, P_{F2}, ..., P_{FN}).$$

Assuming continuity of the probability density function of the likelihood ratios at the local detectors, one can compute the gradient of $\Re(\Gamma)$ with respect to P_{F_i}, $i = 1, ..., N$. Then the steepest descent algorithm or the conjugate gradient method can be employed for system design. Let $F^{(n)}$ be the set of false alarm probabilities at iteration n, $G^{(n)}$ be the corresponding gradient evaluated at $F^{(n)}$ and $d^{(n)}$ be the search direction at iteration n. Then the family of gradient-based algorithms is given by

$$F^{(n+1)} = F^{(n)} + \alpha_n d^{(n)}, \qquad (3.4.39)$$

where α_n, the optimal step length obtained via a line search, is given by

$$\alpha_n = \arg\min_{\alpha \in (0,\infty)} \Re(F^{(n)} + \alpha d^{(n)}). \qquad (3.4.40)$$

For the steepest descent algorithm, $d^{(n)} = -G^{(n)}$. For the conjugate gradient algorithm with Polak–Ribiere update method with restarts, $d^{(n)}$ is given by

$$d^{(n)} = -G^{(n)} + \beta_{n-1} d^{(n-1)}, \qquad (3.4.41)$$

where

$$\beta_{n-1} = \frac{(G^{(n)} - G^{(n-1)})^T G^{(n)}}{(G^{(n)})^T G^{(n)}}.$$

These gradient based iterative algorithms are run until the difference between the Bayesian risk at the end of two consecutive iterations, i.e., $|\Re^{(n+1)} - \Re^{(n)}|$, is less than a prespecified tolerance. Another efficient gradient algorithm for solving this problem has been proposed by Helstrom [Hel95b]. He considers the more general problem of determining quantization levels at the local detectors.

A heuristic algorithm that combines the features of the person-by-person optimization algorithm and the gradient based line search techniques has also been suggested [Tan90]. The Bayesian risk function

$\Re(\Gamma)$ is again considered to be a function of N false alarm probabilities P_{Fi}, $i = 1, ..., N$. The basic idea of the algorithm is to determine the PBPO decision rule for each local detector and seek the maximum improvement in $\Re(\Gamma)$ along the direction from the current strategy to the new strategy implied by the PBPO strategy. Tang implemented the four algorithms proposed by him [Tan90]. Based on the examples considered, it was found that, in terms of computation time, the person-by-person algorithm was most efficient. But it was more sensitive to initial conditions and often converged to different suboptimal solutions for large N. The other three algorithms were found to be more robust to initial conditions. Among these three algorithms, the heuristic algorithm was found to be the most efficient, in general.

Another Formulation

In the above formulation, the parallel fusion network was optimized by considering the minimization of the Bayesian risk function directly. Next, we consider an alternate approach based on the expression for the minimum average cost of detector operation given in (2.2.25).

In the distributed detection network under consideration, the global decision u_0 is made by the fusion center based on the incoming local decision vector $\boldsymbol{u}$. The fusion rule is implemented at the fusion center by optimally partitioning the space consisting of all possible values of $\boldsymbol{u}$ into two mutually exclusive decision regions corresponding to H_0 and H_1, respectively. From (2.2.27), the minimum average cost for decision making at the fusion center for this discrete observation case is given as

$$\Re_{\min} = C_0 - \tfrac{1}{2} \sum_{\boldsymbol{u}} |(C_{01} - C_{11})P_1 p(\boldsymbol{u}|H_1) - (C_{10} - C_{00})P_0 p(\boldsymbol{u}|H_0)|,$$

(3.4.42)

where the notation is as defined before. Again, we consider the case of conditionally independent and identically distributed observations at the local detectors. Also, the local decision rules are restricted to be identical. The local decision rules are threshold tests characterized by a single threshold t. Each local decision u_i can take two possible values, namely, 0 or 1. It can be modeled as a Bernoulli random variable with parameter p_F when H_0 is true and with parameter p_D when H_1 is true. Thus, the vector $\boldsymbol{u}$ contains the results of a sequence of Bernoulli trials.

If B represents the number of ones in u, it has the following conditional distributions:

$$P(B = b|H_0) = \binom{N}{b} p_F^b (1-p_F)^{N-b}, \qquad (3.4.43)$$

and

$$P(B = b|H_1) = \binom{N}{b} p_D^b (1-p_D)^{N-b}. \qquad (3.4.44)$$

Using these in equation (3.4.42), the minimum average cost for the system may be expressed as

$$\mathfrak{R}_{min}(N,t) = C_0 - \tfrac{1}{2} \sum_{b=0}^{N} \binom{N}{b} \big| (C_{01} - C_{11}) P_1 [p_D(t)]^b [1-p_D(t)]^{N-b}$$

$$- (C_{10} - C_{00}) P_0 [p_F(t)]^b [1-p_F(t)]^{N-b} \big|. \qquad (3.4.45)$$

Note that the minimum risk $\mathfrak{R}_{min}$ is a function of the number of local detectors N and the local threshold t. The dependence on t is explicitly shown in (3.4.45). For any given value of the local threshold t, (3.4.45) specifies the minimum achievable average system cost using the best possible fusion rule among all possible fusion rules. For a different value of the local threshold, (3.4.45) specifies the minimum achievable average cost for that local threshold value using, possibly, another fusion rule. Thus, the optimum system can be designed by selecting the value of the local threshold that minimizes $\mathfrak{R}_{min}(N, t)$ and, then, by determining the fusion rule that must be used to attain the minimum value of $\mathfrak{R}_{min}(N, t)$. The fusion rule can be determined from (3.4.24) by using the value of t and considering all possible realizations of u. This design procedure requires minimizing a single function of one variable. Efficient optimization algorithms [Ber82] can be employed for this purpose. A more detailed discussion is available in [Has91].

Parallel Fusion Network with Soft Decisions

The discussion above on the design of parallel fusion networks has been limited to the case of hard decisions where the observation space at each local detector was partitioned into two regions and a binary decision was transmitted from the local detectors to the fusion center. In the more general system, where soft decisions are transmitted from the local detectors to the fusion center, the observation space of local detectors is partitioned into more than two regions. Decision rules to accomplish this partitioning need to be designed. An approach to the design of local decision rules, based on the expression for the minimum average cost of detector operation given in (2.2.25), that is similar to the hard decision case can be followed. Consider the special case in which the observations at the local detectors are conditionally independent. In this case, the local decision rules can be specified in terms of $(J - 1)$ thresholds at each detector where J is the number of partitions at each local detector. Thus, a total of $N(J - 1)$ thresholds need to be determined. Furthermore, if the observations at the local detectors are assumed to be identically distributed and the decision rules are constrained to be identical, only $(J - 1)$ local thresholds for the entire system need to be computed. One approach to evaluate these $(J - 1)$ thresholds is to employ the formulation described above by using the expression for $\Re_{min}$ given in (3.4.45). The decision of local detector DM i can take one of J possible values, i.e., $u_i = j, j = 0, ..., J - 1$. The probabilities of these events are determined by the hypothesis present. Let α_{ji} and β_{ji} be as defined in (3.3.12) and (3.3.13). Because local decision rules are constrained to be identical, $\alpha_{j1} = \alpha_{j2} = \cdots = \alpha_{jN}$ and $\beta_{j1} = \beta_{j2} = \cdots = \beta_{jN}$, $j = 0, 1, ..., J - 1$. Let α_j and β_j represent these two probabilities, i.e.,

$$\alpha_j = P(u_i = j | H_0), \quad j = 0, 1, ..., J-1, \qquad (3.4.46)$$

and

$$\beta_j = P(u_i = j | H_1), \quad j = 0, 1, ..., J-1. \qquad (3.4.47)$$

The elements of the vector u take integer values between 0 and $J - 1$. Let $B_j, j = 0, ..., J - 1$, represent the number of u_i in u that are equal to j. Then,

$$P(B_0 = b_0, ..., B_{J-1} = b_{J-1} | H_0) = \frac{N!}{b_0! ... b_{J-1}!} (\alpha_0)^{b_0} ... (\alpha_{J-1})^{b_{J-1}},$$
(3.4.48)

and

$$P(B_0 = b_0, ..., B_{J-1} = b_{J-1} | H_1) = \frac{N!}{b_0! ... b_{J-1}!} (\beta_0)^{b_0} ... (\beta_{J-1})^{b_{J-1}}.$$
(3.4.49)

Substituting the above probabilities in (3.4.45), the minimum average cost for the system can be expressed as

$$\mathfrak{R}_{min}(N, t_1, ..., t_{J-1}) = C_0 - \tfrac{1}{2} \sum_{b_0} ... \sum_{b_{J-1}} \frac{N!}{b_0! ... b_{J-1}!}$$
$$\times |(C_{01} - C_{11}) P_1 (\beta_0)^{b_0} ... (\beta_{J-1})^{b_{J-1}}$$
$$- (C_{10} - C_{00}) P_0 (\alpha_0)^{b_0} ... (\alpha_{J-1})^{b_{J-1}}|,$$
(3.4.50)

where summations are carried out over all values of $b_0, ..., b_{J-1}$ such that $b_0 + ... + b_{J-1} = N$. The minimum risk is a function of the number of detectors and the values of local thresholds. Optimization algorithms can be employed to determine the values of local thresholds that minimize $\mathfrak{R}_{min}(.)$. Then, as before, the fusion rule yielding the minimum value of $\mathfrak{R}_{min}(.)$ can be determined. The case in which direct observations are available at the fusion center can be handled similarly.

The design of parallel fusion networks employing soft decisions has also been considered by Lee and Chao [LeC89] and by Longo et al. [LLG90]. Both of these design approaches result in systems that perform better than the optimum hard decision systems but do not necessarily result in optimum soft decision systems. The emphasis in [LeC89] and [LLG90] is on tractable and efficient approaches that yield optimal or near optimal systems. Lee and Chao [LeC89] assume that the local detectors first partition their observation space into two mutually exclusive regions to correspond to the two hypotheses. To convey additional information about their confidence in their decision to the fusion center, each of the two mutually exclusive regions is further subpartitioned. Information regarding the subpartition in which the observation lies is conveyed to the fusion center for its use while making the global decision. Subpartitioning is done based on the maximization of the J-divergence. Specifically, the fusion rule (3.3.15) is used. This fusion rule may be expressed as

$$\sum_{i=1}^{N} w_i \underset{u_0=0}{\overset{u_0=1}{\gtrless}} \log \eta, \qquad (3.4.51)$$

where

$$w_i = \begin{cases} \log \dfrac{\beta_{1i}}{\alpha_{1i}}, & \text{if } u_i = H_1, \\ \log \dfrac{\beta_{0i}}{\alpha_{0i}}, & \text{if } u_i = H_0. \end{cases}$$

The J-divergence for the fusion center input is given by

$$\begin{aligned} JD &= E_1 \left(\sum_{i=1}^{N} w_i \right) - E_0 \left(\sum_{i=1}^{N} w_i \right) \\ &= \sum_{i=1}^{N} \left(E_1(w_i) - E_0(w_i) \right) \\ &= \sum_{i=1}^{N} JD_i, \end{aligned} \qquad (3.4.52)$$

where $E_j(\cdot)$ denotes the expectation under hypothesis H_j. JD can be maximized by maximizing each JD_i independently of others. Partitioning of the local detector observation space is obtained by determining the optimum values of α_{ji} and β_{ji}. These are obtained by setting the derivatives of JD_i with respect to these parameters equal to zero and, then, solving the resulting set of equations. In general, this approach does not result in an optimal solution. The suboptimality of this approach stems from the fact that the partitioning of the observation space at the local detectors is carried out in two stages and the fact that the optimization criterion is the maximization of J-divergence which does not minimize the system average cost directly. It is, however, shown that the soft decision system designed using this approach significantly outperforms the optimum system employing hard decisions alone.

Longo, Lookabaugh and Gray [LLG90] view designing the parallel fusion network as a quantization problem where local detectors quantize their observations and transmit them to the fusion center for making the final decision. Even though local detectors quantize their observations separately, they have a common goal, namely, the maximization of a systemwide performance measure. Ali–Silvey distances (also known as f-divergences) have been employed as their optimization criteria. A design algorithm based on cyclic optimization, wherein each optimization step is an alternating minimization based on Lloyd's minimum distortion quantizer design [Llo92], has been proposed. Simplifications to the algorithm to reduce the complexity have also been considered. A number of examples to illustrate the algorithms have been presented. These examples serve to demonstrate the trade-offs of system performance with the number of quantization levels. For details about the algorithm, the reader is referred to [LLG90].

Dependent Observations

One of the major assumptions in this section was that the observations at the local detectors were conditionally independent. Under this assumption, the optimal decision rules are threshold strategies determined by means of a search over a much smaller set of strategies. When the assumption of conditional independence is relaxed, the optimal decision rules need not belong to the set of threshold strategies. This results in a significant increase in the computational complexity of the distributed detection problem. Consider the discrete version of a minimum probability of error distributed detection system without the conditional independence assumption. For a binary hypothesis testing problem using two sensors, assume that local detectors transmit hard decisions and that the fusion center does not receive any direct observations of the phenomenon. Also, let the observations at the two sensors be discrete that belong to finite sets. For this problem, Tsitsiklis and Athans [TsA85] have obtained a fundamental result showing that the probability of deciding whether a decision strategy exists, for which the Bayes risk $\Re$ is less than a given rational number, is *NP*-complete (see [GaJ79] for an introductory treatment of *NP*-completeness). Thus, unless $P = NP$, there is no polynomial time algorithm for this problem. It has been further shown in [PaT86] that similar results regarding the computational complexity of the continuous version of distributed detection problems exist.

Due to the above computational intractability, not much progress on this problem has been made. A major complication is the nonoptimality of threshold strategies. As a result, the parameterization of decision rules is infinite-dimensional. One may obtain suboptimal decision rules by constraining them to a set that admits finite-dimensional parameterization. For example, one may restrict attention to the set of threshold strategies and employ algorithms, such as the ones described earlier, to determine the best solution from this restricted set. The resulting system may yield acceptable performance. This approach has been adopted by Lauer and Sandell [LaS82] where they have considered detection of known and unknown signals in correlated noise. Aalo and Viswanathan [AaV89] have studied the effect of correlated noise on the performance of distributed detection systems. They considered two specific examples, namely, correlated Gaussian and Laplacian noise. Tang [Tan90] has presented a Gauss–Seidel algorithm for the correlated observation case where certain dependencies are neglected for speed up. It is shown that, if this algorithm converges, it yields results that are close to optimal. Development of near optimal detection algorithms appears to be the only practical approach due to the *NP*-completeness of the problem.

Uncertain Observation Models and Robust Detection

In the discussion thus far, we have assumed that the conditional probability distribution of observations under each hypothesis is completely known a priori. In many practical situations, this assumption is not realistic and only partial or incomplete knowledge may be available. Probability distribution may be known approximately or it may be known to belong to certain classes of distributions (uncertainty classes). For example, a reasonably accurate nominal model may be available, but some small deviations from this model may occur. A specific example of this type is the ε-contaminated mixture. In this model, the actual conditional distribution corresponding to H_j is not exactly P_j, $j = 0, 1$, but is given by $(1-\varepsilon)P_j + \varepsilon M_j$, $j = 0, 1$, where P_j are the nominal distributions, M_j are unknown "contaminating" distributions and ε is a number between 0 and 1 that denotes the degree of uncertainty to be included in the model. One approach for designing decision rules in this case is to adopt the philosophy of minimax robustness and design decision rules for the least favorable case (worst case) over the uncertainty classes. Decision rules so designed are said to be robust over the uncertainty classes. Because the robust scheme guarantees a desired level of performance under worst-case conditions, this decision rule will perform better under all other situations within the uncertainty class. Here, we briefly consider robust tests for the distributed detection problem. To provide some background, robust detection for the centralized problem is discussed first.

Consider the binary hypothesis testing problem in the Bayesian framework. System objective is to minimize the probability of error. Let P_0 and P_1 be the true conditional distributions under H_0 and H_1. When P_0 and P_1 are completely known, the decision rule $\gamma(\cdot)$ is designed that minimizes

$$\Re = P_0 P_F(\gamma, P_0) + P_1 P_M(\gamma, P_1), \qquad (3.4.53)$$

where P_0 and P_1 denote the prior probabilities and $P_F(\cdot,\cdot)$ and $P_M(\cdot,\cdot)$ represent probabilities of false alarm and miss. Let ψ_j denote the uncertainty class under H_j, $j = 0, 1$, i.e., P_j lies in the set of distributions ψ_j. Define the worst-case values of the probabilities of false alarm and miss as

$$P_F(\gamma, \psi_0) = \sup_{P_0 \in \psi_0} P_F(\gamma, P_0),$$

and

$$P_M(\gamma, \psi_1) = \sup_{P_1 \in \psi_1} P_M(\gamma, P_1).$$

Now consider the design of decision rule $\gamma_R(\cdot)$ that minimizes the risk function obtained by replacing $P_F(\gamma, P_0)$ and $P_M(\gamma, P_1)$ by $P_F(\gamma, \psi_0)$ and $P_M(\gamma, \psi_1)$. This risk function is given by

$$\Re_R = P_0 P_F(\gamma, \psi_0) + P_1 P_M(\gamma, \psi_1) \qquad (3.4.54)$$

and is based on the worst-case values of the probabilities of false alarm and miss. The resulting decision rule $\gamma_R(\cdot)$ is the desired robust decision rule. It yields the best worst-case performance and provides an upper bound on system performance over uncertainty classes ψ_0 and ψ_1. It is pointed out in [Poo88] that the solution to this problem is obtained by determining the optimum decision rule for the exact model when $\wp_0$ and $\wp_1$ are replaced by a pair $Q_0 \in \psi_0$ and $Q_1 \in \psi_1$ of least favorable distributions.

The same design philosophy of minimax robustness has been employed by Geraniotis and Chau [GeC90] and Veeravalli, Basar and Poor [VBP94a] to obtain robust decision rules for the parallel fusion network. Consider the network shown in Figure 3.6 consisting of N local detectors and a fusion center. The observations y_i, $i = 1, ..., N$, are assumed to be conditionally independent. Let $\wp_0^i$ and $\wp_1^i$, $i = 1, ..., N$, be the true conditional distributions under H_0 and H_1 at the ith detector and let ψ_0^i and ψ_1^i denote the corresponding uncertainty classes. Let $\wp_j$ and ψ_j, $j = 0, 1$, define the product distributions $\wp_j = \wp_j^1 \times ... \times \wp_j^N$ and $\psi_j = \psi_j^1 \times ... \times \psi_j^N$, $j = 0, 1$. As before, $\Gamma = \{\gamma_0, \gamma_1, ..., \gamma_N\}$ represents the set of decision rules at the fusion center and at the local detectors. This set needs to be designed so as to minimize

$$\Re = P_0 P_F(\Gamma, \wp_0) + P_1 P_M(\Gamma, \wp_1). \qquad (3.4.55)$$

Define the worst-case probabilities of false alarm and miss as

$$P_F(\Gamma, \psi_0) = \sup_{\wp_0 \in \psi_0} P_F(\Gamma, \wp_0),$$

and

$$P_M(\Gamma, \psi_1) = \sup_{\wp_1 \in \psi_1} P_M(\Gamma, \wp_1).$$

The set of robust decision rules Γ_R is obtained by minimizing

$$\Re_R = P_0 P_F(\Gamma, \psi_0) + P_1 P_M(\Gamma, \psi_1). \qquad (3.4.56)$$

It has been shown in [GeC90, VBP94a] that robust decision rules are monotonic LRTs and can be obtained by minimizing $\Re$ given in (3.4.55) where $\wp_0$ and $\wp_1$ are replaced by least favorable distributions. Furthermore, the least favorable distributions are the same as those obtained for the centralized problem, i.e., Q_0 and Q_1 as defined earlier. The decision rules so obtained guarantee a minimum level of acceptable performance within the uncertainty classes.

Nonparametric Detection

We consider uncertain observation models a bit further and discuss distributed nonparametric detection briefly. Robust detection techniques described earlier are applicable when observation statistics are not known exactly but are known approximately. Nonparametric techniques are applicable to situations in which very coarse information about observations is available. In this case, one makes some general assumptions about observation statistics, such as symmetry of the observation probability density functions, continuity of the cumulative distribution functions, etc. Because there are a large number of density functions satisfying these general assumptions, they may vary over a wide range without altering detection performance. Most nonparametric detectors employ the sign or the rank of the observed samples. They are easier to implement when compared to parametric detectors. Nonparametric detectors are less efficient than parametric detectors because they rely on less information.

First, we describe two commonly used nonparametric detectors, namely, the sign detector and the Wilcoxon detector. These two nonparametric detectors will be employed to illustrate distributed nonparametric detection. Let $y_1, y_2, ..., y_L$, denote independent and identically distributed observations. The probability density function of observations y_i, $i = 1, ..., L$, under the null hypothesis H_0 is assumed to be symmetric, i.e., $p(y_i) = p(-y_i)$. The signal to be detected is assumed to be positive. The sign detector bases its decision on the signs or two-level approximation of the input data. It uses a hard limiter to obtain

$$v_i = \begin{cases} +1, & \text{if } y_i \geq 0, \\ -1, & \text{if } y_i < 0, \end{cases} \quad i = 1,...,L.$$

The test statistic is given by the sum of v_i, $i = 1, ..., L$, and the nonrandomized test is a threshold test given by

$$\sum_{i=1}^{L} v_i \underset{H_0}{\overset{H_1}{\underset{<}{>}}} t_{\text{sign}}. \quad (3.4.57)$$

The sign test is based only on the signs of the observations; their actual values are not taken into account. The Wilcoxon detector utilizes this information in terms of ranks. Given the set of observations y_i, $i = 1, ..., L$, they are rank-ordered (in an increasing order) based on their absolute value. Let r_i denote the rank of y_i. Then, the nonrandomized Wilcoxon test is given as

$$\sum_{i=1}^{L} r_i \, \text{sgn}(y_i) \underset{H_0}{\overset{H_1}{\underset{<}{>}}} t_{\text{Wilc}}, \quad (3.4.58)$$

where

$$\text{sgn}(y_i) = \begin{cases} 1, & \text{if } y_i \geq 0, \\ 0, & \text{if } y_i < 0. \end{cases}$$

Because the Wilcoxon test uses more information, it performs better than the sign test, but it involves the computationally intensive operation of ranking.

Consider the parallel fusion network topology consisting of N local detectors DM i, $i = 1, ..., N$, and a fusion center. Each DM i receives independent and identically distributed observations $y_i^1, y_i^2, ..., y_i^L$, $i = 1, ..., N$. It is assumed that each detector receives an equal number of observations. The case of unequal observation sizes can be handled similarly. Based on its observations, each DM i computes and transmits its binary decision u_i to the fusion center for decision combining. One may also consider transmission of multibit decisions in a similar manner. Observations at the local detectors are assumed to be conditionally independent. The probability density function of observations under H_0 are assumed to be symmetric, and the signal to be detected is positive. The objective is to design a nonparametric detection system that minimizes the average cost of decision making at the fusion center. One needs to determine local decision rules $\gamma_1(\cdot)$, $\gamma_2(\cdot)$, ..., $\gamma_N(\cdot)$ and the fusion rule $\gamma_0(\cdot)$. Here, we consider a subproblem of the overall design problem and obtain local decision rules $\gamma_1(\cdot)$, ..., $\gamma_N(\cdot)$ for fixed fusion rules. Decision rules at the local detectors can be based on nonparametric detectors, such as the sign detector [Han92, HVV90] or the Wilcoxon detector [Nas93, NaT93]. Here we discuss the use of the more general Wilcoxon detector.

At each DM i, first the Wilcoxon statistic (left-hand side of (3.4.58)) is computed based on the set of received observations $y_i^1, ..., y_i^L$. Let w_i denote this statistic. Local decision rules in terms of w_i need to be determined that minimize the average cost of decision making at the fusion center, as given in (3.4.7). PBPO methodology is employed for system optimization. Following steps described earlier in this section, it can be easily shown that the local decision rules are threshold tests given by

$$\frac{P(w_k|H_1)}{P(w_k|H_0)} \overset{u_k=1}{\underset{u_k=0}{\gtrless}} \frac{\sum_{u^k,w^k} C_F A(u^k) \prod_{i=1,i\neq k}^{N} P(u_i|w_i)P(w_i|H_0)}{\sum_{u^k,w^k} C_D A(u^k) \prod_{i=1,i\neq k}^{N} P(u_i|w_i)P(w_i|H_1)},$$

$$k=1,...,N, \qquad (3.4.59)$$

where $w^k = (w_1, ..., w_{k-1}, w_{k+1}, ..., w_N)^T$ and the rest of the terms are as defined earlier.

Example 3.8

Consider the parallel fusion network consisting of two local detectors and a fusion center. Assume that the signal and noise conditions at the

Table 3.6. Performance of Distributed Nonparametric Detection

P_F	P_D Sign Detector	Wilcoxon Detector
0.08	0.718	0.832
0.12	0.787	0.880
0.16	0.832	0.910

two local detectors are identical. The signal is assumed to be constant with the value 0.5. Noise is assumed to be Gaussian with zero mean and unit variance. The costs of correct decisions are assumed to be zero, i.e., $C_{00} = C_{11} = 0$. Other costs C_{01} and C_{10} need not be equal and are varied to obtain different error probabilities. Twelve samples are processed at each detector. The performance of the distributed Wilcoxon detectors is evaluated by numerical computation of Wilcoxon statistics. For comparison purposes, the performance of the distributed sign detector is also presented. For the range of values of P_F considered, the AND fusion rule performs better than the OR fusion rule. Numerical results for the AND fusion rule are shown in Table 3.6 [Nas93]. The performance of the distributed Wilcoxon detector is better and is achieved at the expense of additional computation.

Notes and Suggested Reading

Team decision problems have been discussed in [Rad62, MaR72]. The Bayesian distributed detection problem was first studied in [TeS81a] where the binary hypothesis testing problem for the parallel network without fusion was considered. Optimum fusion rule was derived in [ChV86]. Properties of fusion rules were studied in [TVB87a, TVB89]. Fusion rules for other models, such as asynchronous decisions and correlated decisions, are available in [ChK94, DrL91, KZG92]. Design of the entire parallel fusion network based on the Bayesian formulation was considered in [Hob86, HoV89b]. It is further discussed in [Sri86c, ReN87a, ReN87b, Tan90, TPK91a]. An excellent source for an overview and for a discussion on several important issues, such as randomization and computational complexity, is [Tsi93a]. Some other references that deal with these issues are [TsA85, Tsi88, Pap90, WiW92]. Detailed descriptions of several computational algorithms and their comparative evaluation is available in [Tan90, Hel95b]. Design of soft decision systems is considered in [LeC89, LLG90]. Robust and nonparametric approaches are presented in [GeC90, VBP94a, Han92, Nas93].

4
Distributed Bayesian Detection: Other Network Topologies

4.1 Introduction

Bayesian hypothesis testing for the parallel fusion network was discussed extensively in Chapter 3. This chapter considers the problem of Bayesian hypothesis testing for several other network topologies. In Section 4.2, we consider the serial or tandem network, a widely studied network topology. System design methodology is developed and its performance is compared with that of the parallel network. Interesting issues, such as sequencing and placement of detectors, are also discussed. A brief discussion on tree networks is presented in Section 4.3. In Section 4.4, distributed detection networks with feedback are treated. In this class of networks, information flows both downstream as well as upstream, i.e., toward the fusion center and away from it. Feedback is shown to improve system performance. An important issue for this configuration is its data transmission requirement. Two protocols are presented that reduce this requirement. Finally, a unified methodology is presented to represent any decentralized detection network structure. Decision rules are also obtained. This methodology is applicable to detection networks that include memory as well as feedback.

4.2 The Serial Network

We consider the binary hypothesis testing problem for a serial or tandem detection network. All of the detectors are connected in series and receive direct observations of the common phenomenon. The decision of the first detector is based solely on its observation. This decision is transmitted to the second detector which uses it in conjunction with its direct observation to make its decision that is sent to the next detector. This process is repeated at each of the detectors in the network. The decision of the last detector is accepted as the final decision of the network.

We begin by considering a serial network consisting of two detectors as shown in Figure 4.1. The two hypotheses are again denoted by H_0 and H_1 with associated a priori probabilities P_0 and P_1. The observations at the two detectors are denoted by y_1 and y_2, respectively. The joint conditional density of the observations is given by $p(y_1, y_2|H_k)$, $k=0, 1$. Detector 1 makes the decision u_1 based on its observation y_1. This decision u_1 is furnished to the second detector which makes the final decision u_2 based on u_1 and its observation y_2. Both decisions are binary, i.e.,

$$u_i = \begin{cases} 0, & H_0 \text{ is declared present}, \\ 1, & H_1 \text{ is declared present}. \end{cases} \quad (4.2.1)$$

In this case, u_2 is the global decision. The cost of different courses of action is denoted by C_{jk} which is the cost of deciding $u_2=j$ when H_k is present. The objective is to derive decision rules at both detectors so as to minimize the overall average cost, i.e., the average cost of making the decision u_2. In this cost formulation, the cost of deciding $u_1=i$ when H_k is present, is included in the minimization process in an indirect manner. More general cost formulations can be considered, e.g., the cost of decision making of individual detectors can be included explicitly in the system cost. But, here, we confine our attention to the case in which the

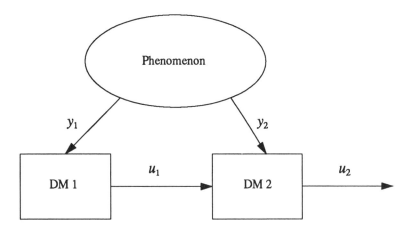

Figure 4.1. A two-detector serial network.

system cost depends only on the final decision of the serial network. We shall assume that the observations y_1 and y_2 are conditionally independent given any hypothesis. Also, the cost of making an incorrect decision is larger than the cost of making a correct decision.

System optimization is carried out based on the PBPO methodology. A decision rule at each detector is derived assuming a fixed decision rule at the other detector. The Bayes risk function for this hypothesis testing problem is given by

$$\begin{aligned}
\Re &= \sum_{i,j,k} \int_{y_1,y_2} p(u_1, u_2, y_1, y_2, H_k) C_{jk} \\
&= \sum_{i,j,k} \int_{y_1,y_2} P_k C_{jk} p(u_1, u_2, y_1, y_2 | H_k) \\
&= \sum_{i,j,k} \int_{y_1,y_2} P_k C_{jk} p(u_2 | u_1, y_1, y_2, H_k) p(u_1, y_1, y_2 | H_k) \ .
\end{aligned}$$

(4.2.2)

Note that the decision u_2 is not a function of y_1 and does not depend on the hypothesis present, i.e., H_k. Also, the observations are conditionally independent. Therefore,

4. Distributed Bayesian Detection: Other Network Topologies

$$\Re = \sum_{i,j,k} \int_{y_1,y_2} P_k C_{jk} p(u_2|u_1, y_2) p(u_1, y_1|H_k) p(y_2|H_k). \quad (4.2.3)$$

Explicitly summing over u_2,

$$\Re = \sum_{i,k} \int_{y_1,y_2} [P_k C_{0k} p(u_2=0|u_1, y_2) p(u_1, y_1|H_k) p(y_2|H_k)$$
$$+ P_k C_{1k} p(u_2=1|u_1, y_2) p(u_1, y_1|H_k) p(y_2|H_k)]. \quad (4.2.4)$$

Setting $p(u_2=1|u_1, y_2) = 1 - p(u_2=0|u_1, y_2)$,

$$\Re = \sum_{i,k} \int_{y_1,y_2} P_k C_{1k} p(u_1, y_1|H_k) p(y_2|H_k)$$
$$+ \sum_i \int_{y_2} p(u_2=0|u_1, y_2) \sum_k \int_{y_1} P_k (C_{0k} - C_{1k}) p(u_1, y_1|H_k) p(y_2|H_k).$$

$$(4.2.5)$$

The first term is constant with respect to u_2 and can be ignored while determining the decision rule at the second detector. The risk $\Re$ is minimized if we set

$$p(u_2=0|u_1, y_2) = \begin{cases} 0, & \text{if } \sum_k \int_{y_1} P_k(C_{0k}-C_{1k}) p(u_1, y_1|H_k) p(y_2|H_k) > 0, \\ 1, & \text{otherwise}. \end{cases} \quad (4.2.6)$$

Integrating over y_1 and expanding over H_k, the condition for making the decision $u_2=1$ becomes

$$\sum_k P_k(C_{0k}-C_{1k}) p(u_1|H_k) p(y_2|H_k) > 0,$$

or

$$P_0(C_{00}-C_{10})p(u_1|H_0)p(y_2|H_0)+P_1(C_{01}-C_{11})p(u_1|H_1)p(y_2|H_1)>0 \;,$$

or

$$P_1(C_{01}-C_{11})p(u_1|H_1)p(y_2|H_1)-P_0(C_{10}-C_{00})p(u_1|H_0)p(y_2|H_0)>0 \;,$$

or

$$\frac{p(y_2|H_1)}{p(y_2|H_0)} > \frac{P_0(C_{10}-C_{00})}{P_1(C_{01}-C_{11})} \frac{p(u_1|H_0)}{p(u_1|H_1)} \;. \tag{4.2.7}$$

The decision rule at the second detector is, therefore,

$$\frac{p(y_2|H_1)}{p(y_2|H_0)} \overset{u_2=1}{\underset{u_2=0}{\gtrless}} \frac{P_0(C_{10}-C_{00})}{P_1(C_{01}-C_{11})} \frac{p(u_1|H_0)}{p(u_1|H_1)} = \frac{C_F}{C_D} \frac{p(u_1|H_0)}{p(u_1|H_1)} \;,$$

$$\tag{4.2.8}$$

where C_F and C_D are the same as defined earlier. Note that this is a threshold test with two possible threshold values, one when $u_1=1$ is received and the other when $u_1=0$ is received. We denote these thresholds by t_2^1 and t_2^0, respectively. They are given by

$$t_2^1 = \frac{C_F}{C_D} \frac{p(u_1=1|H_0)}{p(u_1=1|H_1)} = \frac{C_F}{C_D} \frac{P_{F1}}{P_{D1}} \;, \tag{4.2.9}$$

and

$$t_2^0 = \frac{C_F}{C_D} \frac{p(u_1=0|H_0)}{p(u_1=0|H_1)} = \frac{C_F}{C_D} \frac{1-P_{F1}}{1-P_{D1}} \;. \tag{4.2.10}$$

Note that P_{F1} and P_{D1} are functions of the local threshold at detector 1, i.e., t_1.

Next, we determine the decision rule at the first detector. The risk $\Re$, as given before, is expressed by

$$\Re = \sum_{i,j,k} \int_{y_1,y_2} P_k C_{jk} p(u_2|u_1, y_2) p(u_1, y_1|H_k) p(y_2|H_k) \ . \quad (4.2.11)$$

Decision u_1 is based only on y_1 and does not depend on H_k. Using this fact, we may write

$$\Re = \sum_{i,j,k} \int_{y_1,y_2} P_k C_{jk} p(u_2|u_1, y_2) p(u_1|y_1) p(y_1|H_k) p(y_2|H_k) \ .$$

(4.2.12)

Expanding over u_1,

$$\Re = \sum_{j,k} \int_{y_1,y_2} P_k C_{jk} \big[p(u_2|u_1=0, y_2) p(u_1=0|y_1)$$
$$+ p(u_2|u_1=1, y_2) p(u_1=1|y_1) \big] p(y_1|H_k) p(y_2|H_k) \ .$$

(4.2.13)

Setting $p(u_1=1|y_1) = 1-p(u_1=0|y_1)$, we obtain

$$\Re = \sum_{j,k} \int_{y_1,y_2} P_k C_{jk} p(u_2|u_1=1, y_2) p(y_1|H_k) p(y_2|H_k)$$
$$+ \int_{y_1} p(u_1=0|y_1) \sum_{j,k} \int_{y_2} P_k C_{jk} p(y_1|H_k) p(y_2|H_k)$$
$$\times \big[p(u_2|u_1=0, y_2) - p(u_2|u_1=1, y_2) \big] \ . \quad (4.2.14)$$

The first term is constant with respect to decision making at detector 1 and can be ignored while deriving the decision rule at detector 1. The risk $\Re$ is minimized if we set

$$p(u_1=0|y_1) = \begin{cases} 0, & \text{if } \sum_{j,k} \int_{y_2} P_k C_{jk} p(y_1|H_k) p(y_2|H_k) \\ & [p(u_2|u_1=0, y_2) - p(u_2|u_1=1, y_2)] > 0, \\ 1, & \text{otherwise}. \end{cases} \quad (4.2.15)$$

Integrating over y_2, the decision rule at detector 1 can be expressed as

$$\sum_{j,k} P_k C_{jk}\, p(y_1|H_k)[p(u_2|u_1=0, H_k) - p(u_2|u_1=1, H_k)] \underset{u_1=0}{\overset{u_1=1}{\gtrless}} 0.$$

(4.2.16)

Expanding over j and k,

$$P_0 p(y_1|H_0) \begin{Bmatrix} C_{00}[p(u_2=0|u_1=0, H_0) - p(u_2=0|u_1=1, H_0)] \\ + C_{10}[p(u_2=1|u_1=0, H_0) - p(u_2=1|u_1=1, H_0)] \end{Bmatrix}$$

$$+ P_1 p(y_1|H_1) \begin{Bmatrix} C_{01}[p(u_2=0|u_1=0, H_1) - p(u_2=0|u_1=1, H_1)] \\ + C_{11}[p(u_2=1|u_1=0, H_1) - p(u_2=1|u_1=1, H_1)] \end{Bmatrix} \underset{u_1=0}{\overset{u_1=1}{\gtrless}} 0.$$

(4.2.17)

Setting $p(u_2=0|u_1=i, H_j) = 1 - p(u_2=1|u_1=i, H_j)$, $i, j=0, 1$, and collecting terms, we obtain

$$P_1 \, p(y_1|H_1)\{(C_{01}-C_{11})[p(u_2=1|u_1=1, H_1)-p(u_2=1|u_1=0, H_1)]\}$$

$$-P_0 \, p(y_1|H_0)\{(C_{10}-C_{00})[p(u_2=1|u_1=1, H_0)-p(u_2=1|u_1=0, H_0)]\}$$

$$\begin{array}{c} u_1=1 \\ > \\ < \\ u_1=0 \end{array} 0 . \qquad (4.2.18)$$

As assumed before, $C_{01} > C_{11}$ and $C_{10} > C_{00}$. We also assume that $p(u_2=1|u_1=1, H_k) > p(u_2=1|u_1=0, H_k)$, $k=0, 1$. In other words, we assume that the probability of deciding $u_2=1$ is higher when the incoming decision is $u_1=1$ as opposed to $u_1=0$. Therefore, the coefficients of $p(y_1|H_1)$ and $p(y_1|H_0)$ in (4.2.18) are positive. The decision rule, therefore, is expressed as

$$\frac{p(y_1|H_1)}{p(y_1|H_0)} \begin{array}{c} u_1=1 \\ > \\ < \\ u_1=0 \end{array} \frac{P_0(C_{10}-C_{00})[p(u_2=1|u_1=1, H_0)-p(u_2=1|u_1=0, H_0)]}{P_1(C_{01}-C_{11})[p(u_2=1|u_1=1, H_1)-p(u_2=1|u_1=0, H_1)]} ,$$

$$(4.2.19)$$

or

$$\frac{p(y_1|H_1)}{p(y_1|H_0)} \begin{array}{c} u_1=1 \\ > \\ < \\ u_1=0 \end{array} \frac{C_F}{C_D} \frac{P_{F2}(t_2^1)-P_{F2}(t_2^0)}{P_{D2}(t_2^1)-P_{D2}(t_2^0)} \triangleq t_1 , \qquad (4.2.20)$$

where $P_{F2}(t_2^j)$ and $P_{D2}(t_2^j)$ represent the values of probabilities of false alarm and detection at detector 2 based on threshold value t_2^j, $j=0, 1$. Thus, the threshold t_1 depends explicitly on both possible values of the threshold t_2.

In the above derivation, while deriving the decision rule at one detector, the decision rule at the other detector was assumed to be fixed. The resulting set of three coupled equations in three unknowns needs to be solved simultaneously to yield the three thresholds. These equations represent necessary conditions, and the resulting solution is a person-by-person optimum solution. Computational algorithms, such as the ones presented in Section 3.4, may be employed to obtain a person-by-person optimum solution. Another interesting formulation for serial networks based on deterministic, multistage, nonlinear, optimal control theory is presented in [Tan90]. Efficient computational algorithms based on this formulation are also presented.

Example 4.1

Consider the two-detector serial network shown in Figure 4.1. The observations at the two detectors are assumed to be conditionally independent and identically distributed. Let the conditional densities be Gaussian with unit variance. Under H_0 and H_1, the means are assumed to be zero and one, respectively. Minimum probability of error cost assignment is employed. The three thresholds t_1, t_2^0 and t_2^1 are obtained by a simultaneous solution of (4.2.9), (4.2.10), and (4.2.20). Resulting threshold values as a function of the prior probability P_0 are shown in Figure 4.2.

For this serial network, P_F and P_D are the same as P_{F2} and P_{D2}. They are given by

$$\begin{aligned} P_F &= P(u_2=1 \mid H_0) \\ &= P(u_2=1 \mid u_1=0, H_0) P(u_1=0 \mid H_0) \\ &\quad + P(u_2=1 \mid u_1=1, H_0) P(u_1=1 \mid H_0) \\ &= P_{F2}(t_2^0)(1-P_{F1}) + P_{F2}(t_2^1) P_{F1} , \end{aligned}$$

and

$$\begin{aligned} P_D &= P(u_2=1 \mid H_1) \\ &= P(u_2=1 \mid u_1=0, H_1) P(u_1=0 \mid H_1) \\ &\quad + P(u_2=1 \mid u_1=1, H_1) P(u_1=1 \mid H_1) \\ &= P_{D2}(t_2^0)(1-P_{D1}) + P_{D2}(t_2^1) P_{D1} . \end{aligned}$$

128 4. Distributed Bayesian Detection: Other Network Topologies

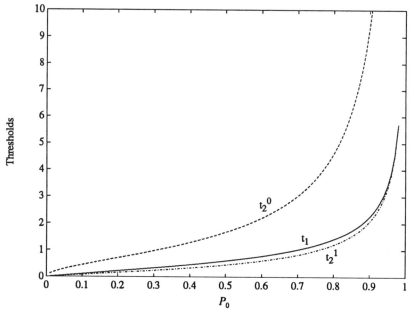

Figure 4.2. Thresholds as a function of P_0 for Example 4.1.

System performance in terms of the average probability of error is obtained as

$$P(e) = P_0 P_F + P_1(1-P_D) .$$

It is shown in Figure 4.3 as a function of the prior probability P_0. Also shown are the corresponding curves for the parallel fusion network using OR and AND fusion rules. Note that, in this case, the serial network performs better than the parallel network.

Tandem networks with more than two detectors can be treated similarly. For example, consider the three detector serial network shown in Figure 4.4. In this case, detector 1 generates u_1 based on its observation y_1. Detector 2 yields decision u_2 based on u_1 and its observation y_2. Finally, detector 3 computes the final decision u_3 based on u_2 and observation y_3. The system can be characterized in terms of five thresholds t_1, t_2^0, t_2^1, t_3^0 and t_3^1 where t_i^j represents the threshold of the ith detector when $u_{i-1}=j$, $i=2, 3, j=0, 1$. The necessary conditions

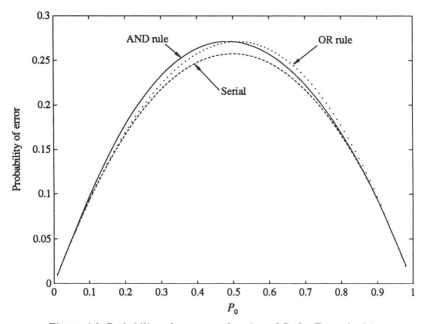

Figure 4.3. Probability of error as a function of P_0 for Example 4.1.

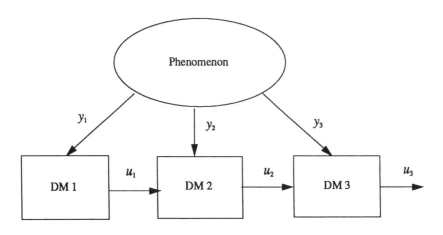

Figure 4.4. Block diagram of a three-sensor serial network.

to obtain the PBPO values of the thresholds can be obtained in a manner similar to that used for the serial network consisting of two detectors. The set of five coupled equations follows:

$$t_1 = \frac{C_F}{C_D} \frac{P_{F2}(t_2^1)-P_{F2}(t_2^0)}{P_{D2}(t_2^1)-P_{D2}(t_2^0)} \frac{P_{F3}(t_3^1)-P_{F3}(t_3^0)}{P_{D3}(t_3^1)-P_{D3}(t_3^0)}, \qquad (4.2.21)$$

$$t_2^0 = \frac{C_F}{C_D} \frac{1-P_{F1}(t_1)}{1-P_{D1}(t_1)} \frac{P_{F3}(t_3^1)-P_{F3}(t_3^0)}{P_{D3}(t_3^1)-P_{D3}(t_3^0)}, \qquad (4.2.22)$$

$$t_2^1 = \frac{C_F}{C_D} \frac{P_{F1}(t_1)}{P_{D1}(t_1)} \frac{P_{F3}(t_3^1)-P_{F3}(t_3^0)}{P_{D3}(t_3^1)-P_{D3}(t_3^0)}, \qquad (4.2.23)$$

$$t_3^0 = \frac{C_F}{C_D} \frac{1-P_{F1}(t_1)P_{F2}(t_2^1)-[1-P_{F1}(t_1)]P_{F2}(t_2^0)}{1-P_{D1}(t_1)P_{D2}(t_2^1)-[1-P_{D1}(t_1)]P_{D2}(t_2^0)}, \qquad (4.2.24)$$

and

$$t_3^1 = \frac{C_F}{C_D} \frac{P_{F1}(t_1)P_{F2}(t_2^1)+[1-P_{F1}(t_1)]P_{F2}(t_2^0)}{P_{D1}(t_1)P_{D2}(t_2^1)+[1-P_{D1}(t_1)]P_{D2}(t_2^0)}. \qquad (4.2.25)$$

These may be solved simultaneously to determine the five thresholds. For a serial network consisting of N detectors, there are $(2N-1)$ thresholds that need to be determined by a simultaneous solution of the following $(2N-1)$ coupled equations:

$$t_1 = \frac{C_F}{C_D} \frac{P_{FN}(u_1=1)-P_{FN}(u_1=0)}{P_{DN}(u_1=1)-P_{DN}(u_1=0)}, \qquad (4.2.26)$$

$$t_n^i = \frac{C_F}{C_D} \frac{P_{FN}(u_n=1) - P_{FN}(u_n=0)}{P_{DN}(u_n=1) - P_{DN}(u_n=0)} \frac{P(u_{n-1}=i|H_0)}{P(u_{n-1}=i|H_1)},$$

$$n = 2, 3, \ldots, N-1;\ i = 0, 1, \tag{4.2.27}$$

$$t_N^i = \frac{C_F}{C_D} \frac{P(u_{N-1}=1|H_0)}{P(u_{N-1}=i|H_1)}, \quad i = 0, 1, \tag{4.2.28}$$

where

$$P_{FN}(u_n=i) = P(u_N=1|u_n=i, H_0),\ i=0,1,$$

and

$$P_{DN}(u_n=i) = P(u_N=1|u_n=i, H_1),\ i=0,1.$$

When the number of detectors in a tandem network becomes larger, the above approach for representing the necessary conditions becomes tedious. A more convenient approach is to represent the above set of equations in a recursive fashion. Let us consider the recursive formulation for the serial network consisting of three detectors. It can be considered a serial network consisting of two processors. The first processor consists of the first two detectors and the second processor is the last detector, i.e., detector 3. In this case, the threshold equations become

$$t_2 = \frac{C_F}{C_D} \frac{P_{F3}(t_3^1) - P_{F3}(t_3^0)}{P_{D3}(t_3^1) - P_{D3}(t_3^0)}, \tag{4.2.29}$$

$$t_3^0 = \frac{C_F}{C_D} \frac{1 - Q_{F2}(t_2)}{1 - Q_{D2}(t_2)}, \tag{4.2.30}$$

and

$$t_3^1 = \frac{C_F}{C_D} \frac{Q_{F2}(t_2)}{Q_{D2}(t_2)}, \qquad (4.2.31)$$

where

$$Q_{F2}(t_2) = P_{F1}(t_1)P_{F2}(t_2^1) + [1-P_{F1}(t_1)]P_{F2}(t_2^0), \qquad (4.2.32)$$

and

$$Q_{D2}(t_2) = P_{D1}(t_1)P_{D2}(t_2^1) + [1-P_{D1}(t_1)]P_{D2}(t_2^0). \qquad (4.2.33)$$

The thresholds t_1, t_2^0, and t_2^1, are as defined in (4.2.9), (4.2.10) and (4.2.20) for the two-detector serial network. Thresholds t_2^0 and t_2^1 can be expressed in terms of t_2 as follows:

$$t_2^0 = \frac{1-P_{F1}(t_1)}{1-P_{D1}(t_1)} t_2, \qquad (4.2.34)$$

and

$$t_2^1 = \frac{P_{F1}(t_1)}{P_{D1}(t_1)} t_2. \qquad (4.2.35)$$

This recursive procedure can be generalized to a serial network consisting of N detectors. Details are left as an exercise for the reader.

Example 4.2

Consider the serial network shown in Figure 4.4. The same observation statistics and cost assignment as Example 4.1 are assumed. Thresholds are computed by a simultaneous solution of (4.2.21) to (4.2.25). The average probability of error as a function of P_0 is shown in Figure 4.5. The performance of the corresponding parallel fusion network using OR, AND, and MAJORITY fusion rules is also shown. In this case, the serial network performs better than the parallel network using OR and AND fusion rules. The performance of the MAJORITY fusion rule is practically the same as the serial network.

In Examples 4.1 and 4.2, we have compared the performances of the parallel and the serial networks with two and three detectors for Gaussian observation statistics. An interesting question arises as to whether the parallel topology or the serial topology performs better, in general. For the case of two detectors, the following definitive statement can be made [Pap90, Tsi93a]:

Proposition 4.1

For distributed detection networks consisting of two detectors, the tandem network performs at least as well as the parallel network.

Proof

Consider a parallel fusion network with two local detectors and a fusion center. Let $\Gamma^* = \{\gamma_0^*, \gamma_1^*, \gamma_2^*\}$ be the set of optimal decision rules for the fusion center and the two local detectors. Decision rules γ_1^* and γ_2^* operate exclusively on their observations y_1 and y_2 to yield the decisions

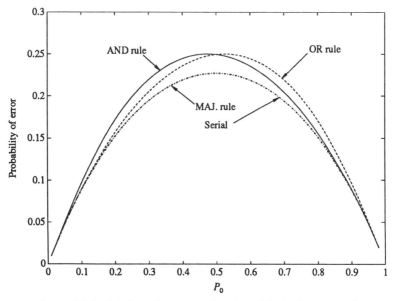

Figure 4.5. Probability of error as a function of P_0 for Example 4.2.

u_1 and u_2. The fusion rule γ_0^* determines the global decision u_0 based on the local decisions u_1 and u_2.

Now, consider a two-detector tandem network in which the detectors employ the decision rules γ_1^* and γ_2^*. The first detector employs γ_1^* to operate on its observation and provides its decision to the second detector. The second detector employs γ_2^* to operate on its observation to come up with its preliminary decision. Then, it uses the fusion rule γ_0^* to combine its preliminary decision and the decision received from the first detector to yield the final decision. The tandem network designed in this ad hoc (not necessarily optimal) manner can always duplicate the performance of the optimal two-detector parallel fusion network. Thus, the tandem network performs at least as well as the parallel network.

The result above does not depend upon the detectors involved, their placement in the tandem configuration, or on prior probabilities and costs. But, for distributed detection networks consisting of more than two detectors, no general results independent of the above factors have been obtained, and different topologies need to be evaluated on a case-by-case basis. In the asymptotic sense, the parallel configuration is superior because, as seen in Chapter 3, its probability of error goes to zero asymptotically. On the other hand, under certain conditions on the likelihood ratio, it has been shown that the asymptotic probability of error for the tandem network is bounded away from zero [PaA92a]. For many probability distributions of interest, e.g., Gaussian and exponential distributions, the probability of error for the serial network can be made to go to zero as $N \to \infty$. However, the convergence rate is slower than exponential. As we have seen, the serial topology is better for $N=2$ and the parallel topology is better for $N \to \infty$. For any given distributed detection problem, a value of N exists at which the parallel topology becomes better. No general results are available that give the value of N at which this transition occurs, but it is conjectured [Tsi93a] that this value is rather small.

Another interesting issue related to the design of serial networks is the ordering or sequencing of detectors in the tandem configuration. Consider a serial network consisting of two detectors one of which is better than the other, i.e., the ROC curve of one detector is higher than that of the other. The problem is to determine the detector to be placed first and the detector to be placed second, i.e., the one that makes the final decision. It seems intuitively appealing to have the better detector make the final decision. But it has been shown by counterexamples in [Pap90] that this ordering is not always optimal. The ordering of detectors depends on other factors, such as prior probabilities and costs. Therefore, no general statements about the sequencing problem can be made, and each situation needs to be investigated to determine the ordering. For exponential distributions, however, it has been shown that the better detector should make the final decision [Pap90].

Example 4.3

Consider a two-detector serial network. The observations at the two detectors are assumed to be conditionally independent but nonidentically distributed. Let all the conditional densities be Gaussian with unit variance. Under H_0 and H_1, the means for the better detector (B) are assumed to be zero and two whereas, for the worse detector (W), they are zero and one. Minimum probability of error cost assignment is employed. System performance in terms of average probability of error versus P_0 is shown in Figure 4.6 for both situations, namely, when W is placed last (that makes the final decision) and when B is placed last. For comparison, system performance, when both detectors are identical and same as W, is also shown. In this case, the performance, when the better detector is placed last, is superior.

Next, consider a three-detector serial network with two detectors the same as W above and one detector the same as B. The system performance when B is placed as the first detector, the middle detector, or the last detector is shown in Figure 4.7. Also shown is the system performance when all the detectors are the same as W. Once again, placement of the better detector as the last detector yields the best performance for this example.

Reibman [Rei87], Tang et al. [TPK91b] and Papastavrou [Pap90] have studied the Bayesian hypothesis testing problem for the serial topology in great detail and have presented extensive numerical results. Papastavrou [Pap90] has also presented two suboptimum schemes, one in which each detector is selfish employing a greedy rule and the second in which all detectors (except the first one) are constrained to employ identical decision rules. Suboptimum systems are much easier to implement, and yet their performance is fairly close to the optimum system.

136   4. Distributed Bayesian Detection: Other Network Topologies

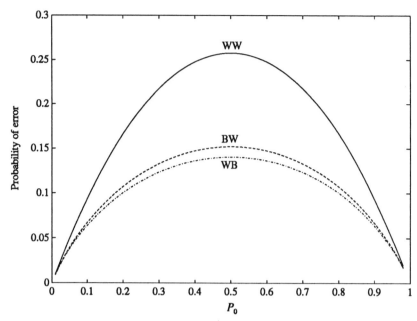

Figure 4.6. System performance for different sequencing of detectors in a two-sensor serial network.

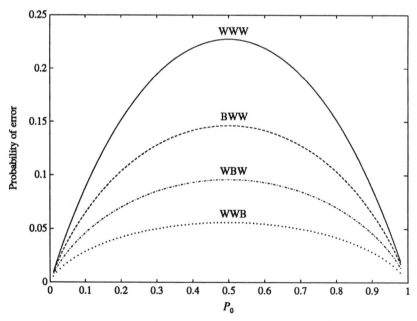

Figure 4.7. System performance for different sequencing of detectors in a three-sensor serial network.

4.3 Tree Networks

Thus far, we have solved the Bayesian hypothesis testing problem for the parallel fusion network and the serial network. As indicated earlier, detection networks can be organized in a variety of configurations. One common example is a tree network. In a tree network, detectors form a directed acyclic graph where the fusion center is the root of the tree and information from all the other detectors flows on a unique path toward the fusion center. In this section, we present an approach to represent tree networks. Bayesian hypothesis testing problem for more general networks that include tree networks as a special case will be solved in Section 4.5.

A detection network is assumed to consist of a number of local detectors and a fusion center responsible for making the global decision. The interconnection structure is specified in terms of the communication matrix D. The elements of this matrix take binary values and indicate the presence or absence of directed communication links between pairs of detectors. Matrix D is of dimension $n \times n$ where n is the total number of detectors in a given system (including the global decision maker).

The (i, k) element of matrix D is equal to one if a directed link exists from the detector corresponding to row i to the detector corresponding to column k. Observe that the kth column of D gives the set of detectors from which the detector corresponding to column k receives its input. We define the decision input of the kth detector as follows:

$$I_k = \{u_i : D_{ik} = 1, \text{ for all } i\} . \quad (4.3.1)$$

Next, we present a few examples to illustrate this representation.

Example 4.4

For a serial system consisting of N detectors, the matrix D is of dimension $N \times N$ and is given by

$$D = \begin{array}{c} \text{det. no.} \\ 1 \\ 2 \\ 3 \\ \vdots \\ N-1 \\ N \end{array} \begin{array}{c} \begin{array}{cccccc} 1 & 2 & 3 & 4 & \cdots & N \end{array} \\ \left(\begin{array}{cccccc} 0 & 1 & 0 & 0 & \cdots & 0 \\ 0 & 0 & 1 & 0 & \cdots & 0 \\ 0 & 0 & 0 & 1 & \cdots & 0 \\ \vdots & \vdots & \cdots & \ddots & \ddots & 0 \\ 0 & 0 & 0 & 0 & \cdots & 1 \\ 0 & 0 & 0 & 0 & \cdots & 0 \end{array} \right) \end{array}$$

The entries of the matrix can be obtained from the block diagram of a serial system similar to Figure 4.4. Detector numbers are also indicated for the convenience of the reader. The (i, k) element is one if detector i transmits its decision to detector k. For example, $D_{12} = 1$ indicates that the decision of detector 1 is fed to detector 2. Using Equation (4.3.1), the decision input of the Nth detector is given by

$$I_N = u_{N-1}.$$

The first column of the D matrix has all zero entries indicating that there is no decision input to detector 1, i.e.,

$$I_1 = \text{no input}.$$

Example 4.5

For the parallel fusion network consisting of N local detectors and a fusion center, the communication matrix D is of dimension $(N+1) \times (N+1)$ and is given by

$$D = \begin{pmatrix} & 1 & 2 & \cdots & N & 0 \\ 1 & 0 & 0 & \cdots & 0 & 1 \\ 2 & 0 & 0 & \cdots & 0 & 1 \\ \vdots & \vdots & \vdots & \vdots & \vdots & \vdots \\ N & 0 & 0 & \cdots & 0 & 1 \\ 0 & 0 & 0 & \cdots & 0 & 0 \end{pmatrix}$$

Note that the global decision maker or the fusion center is denoted by detector number 0. It appears in the last row and the last column of the matrix. As seen from this matrix, there are no decision inputs to the kth detector, $k=1, 2, ..., N$, i.e.,

$$I_k = \text{no input}, \quad k = 1, 2, ..., N.$$

However, the column corresponding to detector 0 (the global decision maker) has the following decision input,

$$I_0 = (u_1, u_2, ..., u_N).$$

Similarly, the communication matrix D for more general tree networks with more degrees of hierarchies can be determined. A generalization of the representation methodology described here will be presented in Section 4.5.

4.4 Detection Networks with Feedback

In the detection network topologies considered thus far, information flowed in only one direction, namely, downstream toward the fusion center. After receiving all of its observations, each local detector made its decision and transmitted it downstream to the next detector with information flowing toward the fusion center. In this section, we consider distributed detection network topologies where information also flows upstream. In these networks, observations arrive sequentially at the local detectors over the observation interval. During this interval, the hypothesis present remains the same. Tentative decisions are made after each observation sample received and are transmitted downstream toward the fusion center. The fusion center and/or detectors along the path to the fusion center combine these incoming decisions with their observations, if any, and transmit their decisions downstream. They may also transmit messages back to the detectors upstream. The detectors adapt their decision rules based on feedback information. In this section, we consider the parallel fusion network topology with feedback. Other network topologies with feedback can be considered similarly.

We consider the binary hypothesis testing problem, with the two hypotheses denoted by H_0 and H_1 respectively, for the system shown in Figure 4.8. This system consists of N local detectors which communicate their decisions to the fusion center after each received observation. The fusion center communicates the global decision back to each of the N detectors after combining each set of received decisions. The system operation is described as follows: At time step t, the kth detector makes the local decision denoted by u_k^t, $k=1, 2, ..., N$, based on the previous global decision denoted by u_0^{t-1}, the current observation denoted by y_k^t and the previous observations y_k^{t-1}, y_k^{t-2}, ..., y_k^1 denoted by $Y_{t-1,k}$. The local decision u_k^t is transmitted to the fusion center where it is combined with the other incoming local decisions to yield the global decision u_0^t. The global decision u_0^t is fed back to all the local detectors for use at the next time step $t+1$.

We assume that the joint conditional probability density functions $p(Y^t, Y^{t-1}, ..., Y^1 | H_j)$, $j=0, 1$, are known a priori where Y^t is the concatenation of all local observations at time step t, i.e., $Y^t = \{y_1^t, y_2^t, ..., y_N^t\}$. The local decision u_k^t is obtained using the decision rule $\gamma_k^t(.)$ as follows:

$$u_k^t = \gamma_k^t(Y_{t,k}, u_0^{t-1}) , \qquad (4.4.1)$$

where $Y_{t,k} = \{y_k^t, y_k^{t-1}, ..., y_k^1\}$. The global decision u_0^t is obtained using the global decision rule $\gamma_0^t(.)$ as follows:

4.4 Detection Networks with Feedback   141

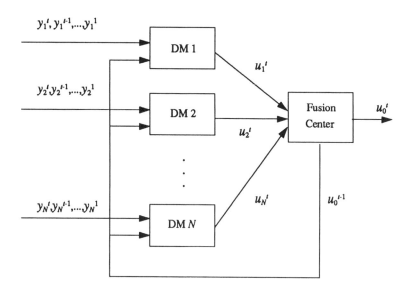

Figure 4.8. Parallel fusion network with feedback.

$$u_0^t = \gamma_0^t(u^t), \qquad (4.4.2)$$

where $u^t = \{u_1^t, u_2^t, ..., u_N^t\}$.

The problem is to find the PBPO decision rules $\gamma_k^t(.)$ for each detector k, $k=0, 1, ..., N$, so as to minimize a given cost function $\Re(\Gamma)$, where

$$\Gamma = \{\Gamma^t : t=1, 2, ...\}$$

and Γ^t is defined as

$$\Gamma^t = \{\gamma_k^t(.) : k=0, 1, ..., N\}.$$

We assume spatial independence, i.e., the observations at the kth detector denoted by $Y_{t,k} = \{y_k^t, y_k^{t-1}, ..., y_k^1\}$ are conditionally independent of the observations at the jth detector, $j \neq k$. Therefore, the a priori knowledge of the conditional probability density functions $p(Y^t, Y^{t-1}, ..., Y^1 | H_j)$, $j=0,1$, reduces to the a priori knowledge of the individual detector conditional probability densities

$p(y_k^t, y_k^{t-1}, ..., y_k^1 | H_j)$, $k=1, 2..., N$; $j=0,1$. In addition, we assume that the observations at the kth detector are conditionally independent in time, i.e., $y_k^t, y_k^{t-1}, ..., y_k^1$ are conditionally independent. Thus, the a priori knowledge of the individual detector conditional probability density reduces further to the knowledge of the conditional probability densities $p(y_k^t | H_j)$, $k=1, 2, ..., N$; $j=0, 1$; $t=1, 2, ...$.

The PBPO decision rules $\gamma_k^t(.)$ for each detector $k=0, 1, ..., N$, are to be found so as to minimize the cost function $\Re(\Gamma)$ given by

$$\Re(\Gamma) = C_{00}P(u_0^t=0, H_0) + C_{01}P(u_0^t=0, H_1)$$
$$+ C_{10}P(u_0^t=1, H_0) + C_{11}P(u_0^t=1, H_1), \quad (4.4.3)$$

where C_{ij}, $i, j=0, 1$, is the cost of deciding $u_0^t=i$ when the true hypothesis is H_j. The costs C_{ij}, $i, j=0, 1$, and the a priori probabilities P_0 and P_1 are assumed to be known. Rewriting (4.4.3) in terms of the probability of false alarm at time step t, P_F^t, and the probability of detection at time step t, P_D^t,

$$\Re(\Gamma) = C_F P_F^t - C_D P_D^t + C, \quad (4.4.4)$$

where C_F, C_D and C are the same as defined in Section 3.4. We first determine the fusion rule $\gamma_0^t(.)$. We expand the probability of false alarm and the probability of detection around the decision vector u^t as follows:

$$\Re(\Gamma) = C_F \sum_{u^t} P(u_0^t=1, u^t | H_0) - C_D \sum_{u^t} P(u_0^t=1, u^t | H_1) + C.$$
$$(4.4.5)$$

Conditioning on u^t and expanding,

$$\Re(\Gamma^t) = C_F \sum_{u^t} P(u_0^t=1|u^t, H_0)P(u^t|H_0)$$
$$- C_D \sum_{u^t} P(u_0^t=1|u^t, H_1)P(u^t|H_1) + C. \quad (4.4.6)$$

Because u_0^t given u^t does not depend on the hypothesis present, we rewrite the previous expression as

$$\Re(\Gamma^t) = \sum_{u^t} P(u_0^t=1|u^t)[C_F P(u^t|H_0) - C_D P(u^t|H_1)] + C. \quad (4.4.7)$$

Due to the PBPO methodology employed, we assume that the local detectors are fixed and minimize the cost function $\Re(\Gamma^t)$ by choosing the decision rule at the fusion center as

$$P(u_0^t=1|u^t) = \begin{cases} 1, & \text{if } C_F P(u^t|H_0) - C_D P(u^t|H_1) < 0, \\ 0, & \text{otherwise}. \end{cases} \quad (4.4.8)$$

This decision rule may be written as

$$\gamma_0^t(u^t) = u_0^t = \begin{cases} 1, & \text{if } \Lambda(u^t) > \dfrac{C_F}{C_D}, \\ 0, & \text{otherwise}, \end{cases} \quad (4.4.9)$$

where

$$\Lambda(u^t) = \frac{P(u^t|H_1)}{P(u^t|H_0)}.$$

Next, we obtain the local decision rules. We write (4.4.4) explicitly, in terms of the kth local decision,

$$\mathcal{R}(\Gamma^t) = \sum_{u_k^t} P(u_0^t=1|u_{k1}^t)[C_F P(u_{k1}^t|H_0) - C_D P(u_{k1}^t|H_1)]$$
$$+ P(u_0^t=1|u_{k0}^t)[C_F P(u_{k0}^t|H_0) - C_D P(u_{k0}^t|H_1)] + C,$$

(4.4.10)

where

$$u_k^t = \{u_1^t, u_2^t, \ldots, u_{k-1}^t, u_{k+1}^t, \ldots, u_N^t\},$$

and

$$u_{ki}^t = \{u_1^t, u_2^t, \ldots, u_k^t=i, \ldots, u_N^t\}, \quad i=0,1.$$

Substituting $P(u_{k0}^t|H_j) = P(u_k^t|H_j) - P(u_{k1}^t|H_j)$, $j=0,1$, in (4.4.10), factoring out common terms, and rearranging,

$$\mathcal{R}(\Gamma^t) = \sum_{u_k^t} [P(u_0^t=1|u_{k1}^t) - P(u_0^t=1|u_{k0}^t)]$$
$$\times [C_F P(u_{k1}^t|H_0) - C_D P(u_{k1}^t|H_1)]$$
$$+ P(u_0^t=1|u_{k0}^t)[C_F P(u_k^t|H_0) - C_D P(u_k^t|H_1)] + C.$$

(4.4.11)

We observe that the last two terms of (4.4.11) are fixed so far as the optimization of the kth local detector is concerned. We ignore these terms in the subsequent equations and denote the remaining terms of the cost function by $\mathcal{R}'(\Gamma^t)$. Expanding $\mathcal{R}'(\Gamma^t)$ in u_0^{t-1}, the previous global decision, and $Y_t = \{Y^t, Y^{t-1}, \ldots, Y^1\}$, the observation vectors of local detectors up to time step t,

$$\Re'(\Gamma^t) = \sum_{u_k^t} [P(u_0^t=1|u_{k1}^t) - P(u_0^t=1|u_{k0}^t)]$$

$$\times \int_{Y_t} \sum_{u_0^{t-1}} [C_F P(u_{k1}^t, u_0^{t-1}, Y_t|H_0) - C_D P(u_{k1}^t, u_0^{t-1}, Y_t|H_1)] \,,$$

(4.4.12)

where $\int_{Y_t}$ is a multifold integral over all y_k^t for all k and all time steps up to and including t. Letting $P(u_0^t=1|u_{k1}^t) - P(u_0^t=1|u_{k0}^t) = f(u_k^t)$ and expanding (4.4.12) by conditioning on u_0^{t-1} and Y_t,

$$\Re'(\Gamma^t) = \sum_{u_k^t} f(u_k^t) \int_{Y_t} \sum_{u_0^{t-1}} [C_F P(u_{k1}^t|u_0^{t-1}, Y_t, H_0)$$

$$\times P(u_0^{t-1}, Y_t|H_0) - C_D P(u_{k1}^t|u_0^{t-1}, Y_t, H_1) P(u_0^{t-1}, Y_t|H_1)].$$

(4.4.13)

The local decision vector u_{k1}^t given both the previous global decision u_0^{t-1} and the observation vector Y_t does not depend on the hypothesis present. In addition, expanding $P(u_0^{t-1}, Y_t|H_j)$, $j=0,1$, by conditioning on Y_t,

$$\Re'(\Gamma^t) = \sum_{u_k^t} f(u_k^t)$$

$$\times \int_{Y_t} \sum_{u_0^{t-1}} [C_F P(u_{k1}^t|u_0^{t-1}, Y_t) P(u_0^{t-1}|Y_t, H_0) p(Y^t, Y_{t-1}|H_0)$$

$$- C_D P(u_{k1}^t|u_0^{t-1}, Y_t) P(u_0^{t-1}|Y_t, H_1) p(Y^t, Y_{t-1}|H_1)] \,,$$

(4.4.14)

where we have used the fact that $Y_t = \{Y^t, Y_{t-1}\}$. The previous global decision u_0^{t-1} does not depend on the observation vector Y^t. Furthermore, using the spatial independence of observations, we rewrite Equation (4.4.14) in terms of the individual detector observation vectors $Y_{t,k}$,

$$\Re'(\Gamma') = \sum_{u_k^t} f(u_k^t)$$

$$\times \int_{Y_{t,1}} \cdots \int_{Y_{t,N}} \sum_{u_0^{t-1}} [C_F P(u_{k1}^t | u_0^{t-1}, Y_t) P(u_0^{t-1} | Y_{t-1}, H_0) \prod_{i=1}^{N} p(Y_{t,i} | H_0)$$

$$- C_D P(u_{k1}^t | u_0^{t-1}, Y_t) P(u_0^{t-1} | Y_{t-1}, H_1) \prod_{i=1}^{N} p(Y_{t,i} | H_1)].$$

(4.4.15)

Because the decision of the kth detector u_k^t depends only on its input observation and does not depend on other detector decisions,

$$P(u_{k1}^t | u_0^{t-1}, Y_t) = P(u_k^t = 1 | u_0^{t-1}, Y_t) \prod_{i=1, i \neq k}^{N} P(u_i^t | u_0^{t-1}, Y_t).$$

Furthermore, due to the spatial independence of observations, the above is written as

$$P(u_{k1}^t | u_0^{t-1}, Y_t) = P(u_k^t = 1 | u_0^{t-1}, Y_{t,k}) \prod_{i=1, i \neq k}^{N} P(u_i^t | u_0^{t-1}, Y_{t,i}).$$

(4.4.16)

Substituting this result in (4.4.15), factoring out the kth local decision term and rearranging,

$$\mathfrak{R}'(\Gamma') = \sum_{u_k^t}\sum_{u_0^{t-1}} f(u_k^t) \int_{Y_{t,k}} P(u_k^t=1 | u_0^{t-1}, Y_{t,k})$$

$$\times \int_{Y_{t,1}} \cdots \int_{Y_{t,k-1}} \int_{Y_{t,k+1}} \cdots \int_{Y_{t,N}} \Big[C_F P(u_0^{t-1} | Y_{t-1}, H_0) p(Y_{t,k} | H_0)$$

$$\times \left[\prod_{i=1, i\neq k}^{N} p(Y_{t,i} | H_0) \right] \left[\prod_{i=1, i\neq k}^{N} P(u_i^t | u_0^{t-1}, Y_{t,i}) \right]$$

$$- C_D P(u_0^{t-1} | Y_{t-1}, H_1) p(Y_{t,k} | H_1) \left[\prod_{i=1, i\neq k}^{N} p(Y_{t,i} | H_1) \right]$$

$$\times \left[\prod_{i=1, i\neq k}^{N} P(u_i^t | u_0^{t-1}, Y_{t,i}) \right]. \qquad (4.4.17)$$

Recalling that u_0^{t-1} does not depend on Y_{t-1}, combining the multiplicative terms in (4.4.17) and unconditioning on $Y_{t,i}$,

$$\mathfrak{R}'(\Gamma') = \sum_{u_k^t}\sum_{u_0^{t-1}} f(u_k^t) \int_{Y_{t,k}} P(u_k^t=1 | u_0^{t-1}, Y_{t,k})$$

$$\times \int_{Y_{t,1}} \cdots \int_{Y_{t,k-1}} \int_{Y_{t,k+1}} \cdots \int_{Y_{t,N}} \Big[C_F P(u_0^{t-1} | H_0) p(Y_{t,k} | H_0)$$

$$\times \left[\prod_{i=1, i\neq k}^{N} p(Y_{t,i}, u_i^t | u_0^{t-1}, H_0) \right] - C_D P(u_0^{t-1} | H_1)$$

$$\times p(Y_{t,k} | H_1) \prod_{i=1, i\neq k}^{N} p(Y_{t,i}, u_i^t | u_0^{t-1}, H_1) \Big]. \qquad (4.4.18)$$

Integrating over $Y_{t,1}, Y_{t,2}, ..., Y_{t,k-1}, Y_{t,k+1}, ..., Y_{t,n}$ and unconditioning on u_0^{t-1}, we rewrite (4.4.18) as

$$\mathfrak{R}'(\Gamma') = \sum_{u_0^{t-1}} \int_{Y_{t,k}} P(u_k^t=1 \,|\, u_0^{t-1}, Y_{t,k}) \sum_{u_k^t} f(u_k^t)$$

$$\times \left[C_F p(Y_{t,k}|H_0) \prod_{i=1, i \neq k}^{N} P(u_i^t, u_0^{t-1}|H_0) - C_D p(Y_{t,k}|H_1) \right.$$

$$\left. \times \prod_{i=1, i \neq k}^{N} P(u_i^t, u_0^{t-1}|H_1) \right]. \qquad (4.4.19)$$

To minimize the cost function given in (4.4.19), we choose

$$P(u_k^t=1 \,|\, u_0^{t-1}, Y_{t,k}) = \begin{cases} 1, & \text{if } A_0 < A_1, \\ 0, & \text{otherwise}, \end{cases} \qquad (4.4.20)$$

where

$$A_1 = \sum_{u_k^t} f(u_k^t) C_D p(Y_{t,k}|H_1) \prod_{i=1, i \neq k}^{N} P(u_i^t, u_0^{t-1}|H_1),$$

and

$$A_0 = \sum_{u_k^t} f(u_k^t) C_F p(Y_{t,k}|H_0) \prod_{i=1, i \neq k}^{N} P(u_i^t, u_0^{t-1}|H_0).$$

The kth detector decision rule is given by rewriting (4.4.20) as

$$\gamma_k^t(Y_{t,k}, u_0^{t-1}) = u_k^t = \begin{cases} 1, & \text{if } \dfrac{p(Y_{t,k}|H_1)}{p(Y_{t,k}|H_0)} > \eta_k^t(u_0^{t-1}), \\ 0, & \text{otherwise}, \end{cases} \qquad (4.4.21)$$

where $\eta_k^t(u_0^{t-1})$ is the threshold of the kth detector at time step t defined as

$$\eta_k^t(u_0^{t-1}) = \frac{C_F \sum_{u_k^t} f(u_k^t) P(u_k^t, u_0^{t-1}|H_0)}{C_D \sum_{u_k^t} f(u_k^t) P(u_k^t, u_0^{t-1}|H_1)} .$$

It is important to observe that the local decision rule is a likelihood ratio test. At time step $t=1$, there is no feedback. At this step, the fusion rule has the same form as given in (4.4.9). However, the local decision rule is a single threshold likelihood ratio test given by

$$\gamma_k^1(y_k^1) = u_k^1 = \begin{cases} 1, & \text{if } \frac{p(y_k^1|H_1)}{p(y_k^1|H_0)} > \eta_k^1, \\ 0, & \text{otherwise}, \end{cases} \quad (4.4.22)$$

where η_k^1 is the kth detector threshold at time step 1 defined as

$$\eta_k^1 = \frac{C_F \sum_{u_k^1} f(u_k^1) P(u_k^1|H_0)}{C_D \sum_{u_k^1} f(u_k^1) P(u_k^1|H_1)},$$

and

$$f(u_k^1) = P(u_0^1 = 1|u_{k1}^1) - P(u_0^1 = 1|u_{k0}^1).$$

For time steps $t>1$, the threshold $\eta_k^t(u_0^{t-1})$ of the kth detector is a function of the previous global decision u_0^{t-1} as shown in Equation (4.4.21). Because the previous global decision u_0^{t-1} may take two values in the case of binary hypothesis testing, two thresholds exist for the likelihood ratio test at the local detectors.

Next, we evaluate the performance of this system in terms of P_F^t and P_M^t and obtain recursive relationships for them. We expand P_F^t defined as

$P(u_0^t = 1 | H_0)$ in terms of u_0^{t-1},

$$P_F^t = P(u_0^t=1|u_0^{t-1}=1, H_0)P(u_0^{t-1}=1|H_0)$$
$$+ P(u_0^t=1|u_0^{t-1}=0, H_0)P(u_0^{t-1}=0|H_0) . \quad (4.4.23)$$

Replacing $P(u_0^{t-1}=0|H_0)$ by $1-P(u_0^{t-1}=1|H_0)$ and rearranging terms,

$$P_F^t = P(u_0^{t-1}=1|H_0)[P(u_0^t=1|u_0^{t-1}=1, H_0)$$
$$-P(u_0^t=1|u_0^{t-1}=0, H_0)]+P(u_0^t=1|u_0^{t-1}=0, H_0). \quad (4.4.24)$$

This may be rewritten as

$$P_F^t = P_F^{t-1}[P_F^t(u_0^{t-1}=1)-P_F^t(u_0^{t-1}=0)]+P_F^t(u_0^{t-1}=0) , \quad (4.4.25)$$

where

$$P_F^t(u_0^{t-1}=i)=P(u_0^t=1|u_0^{t-1}=i, H_0), i=0, 1 .$$

Introducing the local decision vector u^t in the above expression,

$$P_F^t(u_0^{t-1}=i) = P(u_0^t=1|u_0^{t-1}=i, H_0)$$
$$= \sum_{u^t} P(u_0^t=1|u^t, u_0^{t-1}=i, H_0)P(u^t|u_0^{t-1}=i, H_0), i=0, 1.$$

$$(4.4.26)$$

Observing that the global decision u_0^t conditioned on u^t does not depend on u_0^{t-1} and H_0, equation (4.4.26) yields

$$P_F^t(u_0^{t-1}=i)=\sum_{u^t} P(u_0^t=1|u^t)P(u^t|u_0^{t-1}=i, H_0), i=0, 1 . \quad (4.4.27)$$

Similarly, the probability of system miss, P_M^t, is written as

$$P_M^t = P_M^{t-1}[P_M^t(u_0^{t-1}=0) - P_M^t(u_0^{t-1}=1)] + P_M^t(u_0^{t-1}=1) , \qquad (4.4.28)$$

where $P_M^t(u_0^{t-1}=i)$ is expressed as

$$\begin{aligned} P_M^t(u_0^{t-1}=i) &= P(u_0^t=0 | u_0^{t-1}=i, H_1) \\ &= \sum_{u^t} P(u_0^t=0 | u^t) P(u^t | u_0^{t-1}=i, H_1), i=0,1 . \end{aligned}$$

Based on the probabilities of false alarm and miss, we may write the system probability of error as

$$P_e^t = P_F^t P_0 + P_M^t P_1 . \qquad (4.4.29)$$

This characterizes the performance of the overall system.

Introduction of feedback in the parallel fusion network results in performance improvement. Performance superiority of the detection network with feedback over the corresponding network without feedback can be easily established.

Proposition 4.2

The performance of a parallel fusion network with feedback is at least as good as that of the corresponding detection network without feedback.

Proof

Consider a parallel fusion network without feedback and its performance at time t, i.e., performance based on observation samples received up to and including time t. Let $\Gamma^{t*} = \{\gamma_0^{t*}, \gamma_1^{t*}, ..., \gamma_N^{t*}\}$ be the set of optimal decision rules at the fusion center and at the local detectors at time t. Recall that the decision rules at the local detectors are LRTs. Let the thresholds be given by $\eta_1^{t*}, ..., \eta_N^{t*}$. The fusion rule γ_0^{t*} is used to determine u_0^t based on $u_1^t, ..., u_N^t$.

Next, consider the corresponding parallel fusion network with feedback and its performance at time t. In this case also, the decision rules at the local detectors are LRTs that employ two values of thresholds η_k^t ($u_0^{t-1} = 0$) and η_k^t ($u_0^{t-1} = 1$), $k=1, ..., N$. Let us set both values of the threshold to be equal. Specifically,

$$\eta_k^t(u_0^{t-1}=0) = \eta_k^t(u_0^{t-1}=1) = \eta_k^{t*}, \; k=1, ..., N.$$

The fusion rule employed is the same as the network without feedback. The parallel fusion network with feedback, so designed, is able to attain the performance of the optimal parallel fusion network without feedback. Thus, the parallel fusion network with feedback can always be designed to perform as well as the corresponding network without feedback. When it is designed with two values of thresholds that are not necessarily equal and it is optimized, it will perform at least as well as the parallel fusion network without feedback.

Based on the discussion in Chapter 3, the probability of error of the parallel fusion network without feedback goes to zero asymptotically. When feedback is introduced in the network, the probability of error will again go to zero asymptotically and will converge more rapidly.

Data Transmission Protocols

An important issue to be addressed for this detection network with feedback is its data transmission requirements. At each time step, N decisions are transmitted from the local detectors to the fusion center and the global decision is transmitted to N local detectors. It is desirable to reduce this excessive data transmission requirement of $2N$ decision transmissions per time step. This reduction will result in conservation of system resources, such as power and bandwidth. Two protocols that reduce the number of decision transmissions have been proposed in [Alh 90]. In both protocols, the local detectors and the fusion center need to store their previous decision. These protocols rely on the fact that as t increases, detector decisions will begin to agree with their own past decisions and with decisions of other detectors. In this case, decisions need not be transmitted all the time, and, as $t \to \infty$, they need not be transmitted at all. The effectiveness of these protocols can be evaluated in terms of the average number of decision transmissions per time step. This metric consists of two components, namely, decision transmissions downstream (on forward links) and upstream (on feedback links). As SNR and t increase, the average number of decision transmissions per time step decreases quite rapidly. In fact, it is shown in [Alh90] that this metric goes to zero as $t\rightarrow\infty$. Here, we only describe the protocols. Expressions for the metric and the analysis leading to them are available in [Alh90]. Some numerical results to illustrate the performance of the protocols are presented in Example 4.6.

Protocol 1:

In this protocol, at any time step t, the global decision maker transmits its decision to all the local detectors that disagree with it. Therefore, decision transmission on a feedback link takes place only when the global decision maker disagrees with the local detector corresponding to that feedback link, i.e., transmit global decision at time t to local detector k if $\{u_0^t \neq u_k^t, k=1, 2, ..., N\}$. For the forward links, local decision u_k^t is transmitted only if it disagrees with the previous global decision u_0^{t-1}, i.e., transmit local decision at time t from local detector k if $\{u_k^t \neq u_0^{t-1}, k=1, 2, ..., N\}$.

Protocol 2:

In this protocol, at any time step t, the global decision maker transmits its decision to all the local detectors when it disagrees with the previous global decision. Therefore, a feedback decision transmission on all feedback links takes place when the current global decision disagrees with the previous global decision, i.e., transmit global decision at time t to all local detectors if $\{u_0^t \neq u_0^{t-1}\}$. For the forward links, local decision u_k^t is transmitted on the kth forward link only if it disagrees with the previous local decision u_k^{t-1}, i.e., transmit local decision from a local detector k if $\{u_k^t \neq u_k^{t-1}\}$.

Protocol 1 exploits agreements between decisions of local detectors and the fusion center to reduce decision transmissions whereas Protocol 2 takes advantage of agreements between consecutive decisions of local detectors and the fusion center. Distributed detection networks with feedback in conjunction with the above protocols become attractive due to improved performance.

Example 4.6

Consider the parallel fusion network with feedback consisting of two local detectors and a fusion center. Under both hypotheses, input observations at each detector are assumed to follow Gaussian distribution with unit variance. Identical distributions at the two sensors are assumed. Under H_0, the mean is assumed to be zero. Under H_1, the mean is assumed to be S. Minimum probability of error cost assignment and $P_0 = 0.6$ are assumed. Local thresholds at $t=1$ are computed using (4.4.22). The system probability of error is given by (4.4.29), where

$$P_F^1 = \sum_{u^1} P(u_0^1 = 1 | u^1) P(u^1 | H_0),$$

and

$$P_M^1 = \sum_{u^1} P(u_0^1 = 0 | u^1) P(u^1 | H_1).$$

These results are then employed in (4.4.21), (4.4.25), (4.4.28), and (4.4.29) to obtain the thresholds and system probability of error for $t > 1$.

In this example, the fusion rule is not determined using (4.4.9). This fusion rule varies with time because the distributions of u^t vary with time. Instead, fixed fusion rules OR and AND are assumed. Threshold values $\eta_k^t(u_0^{t-1}=0)$ and $\eta_k^t(u_0^{t-1}=1)$ for the two fusion rules are plotted in Figures 4.9 to 4.12 as a function of S. The system probability of error as a function of S for both fusion rules is shown in Figures 4.13 to 4.14. For comparison, system probabilities of error for the parallel fusion network without feedback are given in Figures 4.15 to 4.16.

Observe that, for both fusion rules, $\eta_k^t(u_0^{t-1}=1)$ increases and $\eta_k^t(u_0^{t-1}=0)$ decreases as a function of t. As $t \to \infty$, these thresholds go to zero and one respectively. This indicates asymptotic agreement. For $t=1$, the system probability of error for both systems with and without feedback is the same. For $t > 1$, the parallel fusion network with feedback outperforms the network without feedback.

4.4 Detection Networks with Feedback 155

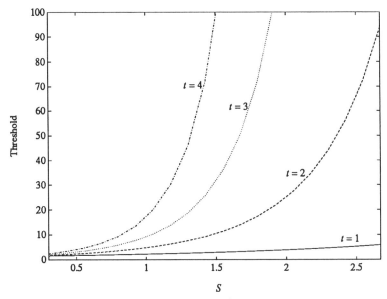

Figure 4.9. Threshold values with $u_0^{t-1} = 0$ for OR fusion rule.

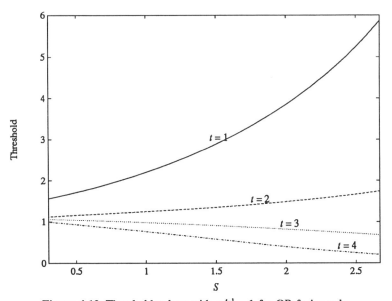

Figure 4.10. Threshold values with $u_0^{t-1} = 1$ for OR fusion rule.

156   4. Distributed Bayesian Detection: Other Network Topologies

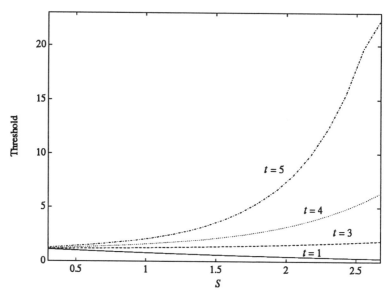

Figure 4.11. Threshold values with $u_0^{t-1} = 0$ for AND fusion rule.

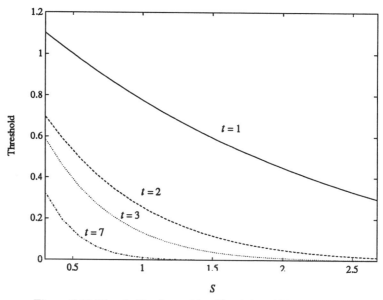

Figure 4.12. Threshold values with $u_0^{t-1} = 1$ for AND fusion rule.

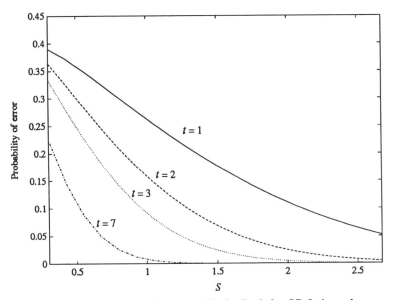

Figure 4.13. System performance with feedback for OR fusion rule.

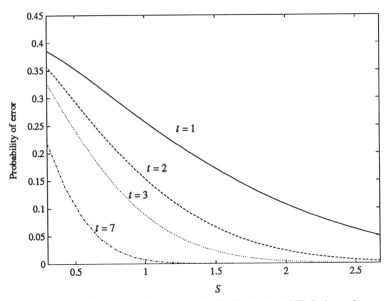

Figure 4.14. System performance with feedback for AND fusion rule.

158   4. Distributed Bayesian Detection: Other Network Topologies

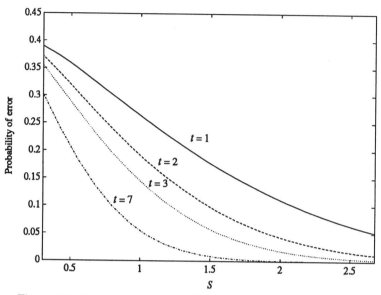

Figure 4.15. System performance without feedback for OR fusion rule.

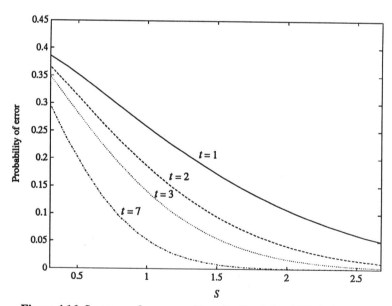

Figure 4.16. System performance without feedback for AND fusion rule.

Next, we present the performance of Protocol 1 in Figures 4.17 to 4.18. The average numbers of decision transmissions per time step as a function of S for OR and AND fusion rules are shown. As anticipated, the average number of decision transmissions decreases as S and t increase. The performance of Protocol 2 is not shown here. It does not perform as well as Protocol 1.

4.5 Generalized Formulation for Detection Networks

In Section 4.3, detection networks organized as directed acyclic graphs were represented by means of a communication matrix D. This representation is not adequate for representing detection networks with more general configurations in which cycles are present, such as the parallel fusion network with feedback considered in Section 4.4. In such systems, different detectors operate on observations collected at different times. Therefore, we generalize the definition of the communication matrix by including a time parameter t. We assume that each detector in a given system produces a unit time delay. Consider the connected graph corresponding to any given decentralized detection network topology, where the nodes represent decision makers and the decisions flow along the directed edges of the graph. Recall the fact that the fusion center is responsible for making the final decision. We organize and label the graph in terms of levels so that the fusion center is at level zero and the level of other nodes is determined by their distance from the fusion center (number of edges traversed from the fusion center to the node under consideration). We illustrate this in Figure 4.19, where a detection network with a non-tree topology along with its corresponding graph is shown. We employ this connected graph to assign time indices to each of the detectors of the detection network. The time index of a detector is simply its level in the connected graph. The time indices of the detectors are displayed along with the detector number in the communication matrix D. Finally, the input decision vector of the detector corresponding to the kth column is given by

$$I_k^t = \left\{ u_i^{t+r_i-c_k-1} : D_{ik} = 1, \text{ for all } i \right\}, \qquad (4.5.1)$$

160   4. Distributed Bayesian Detection: Other Network Topologies

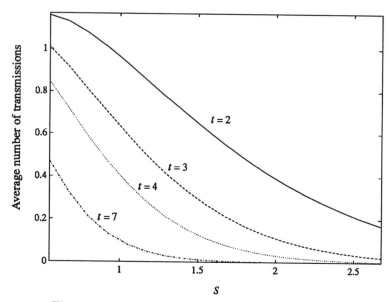

Figure 4.17. Performance of Protocol 1 for OR fusion rule.

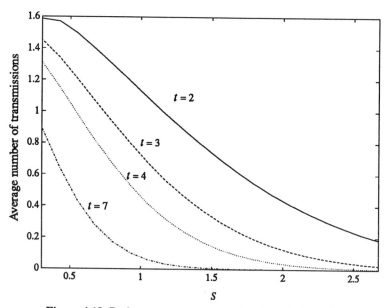

Figure 4.18. Performance of Protocol 1 for AND fusion rule.

4.5 Generalized Formulation for Detection Networks   161

where c_k is the time index of the detector corresponding to the kth column, and r_i is the time index of the detector corresponding to the ith row.

In the decentralized detection system of Figure 4.19, time delays are not shown. Superscript t associated with all observations y_i^t and decisions u_i^t is simply a reminder that proper time indices need to be used during system design and analysis. The corresponding communication matrix is given by

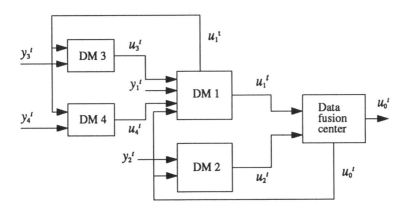

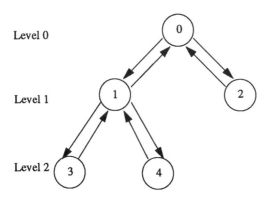

Figure 4.19. A non-tree detection network.

$$D = \begin{matrix} \text{time index} \rightarrow & 1 & 1 & 2 & 2 & 0 \\ \downarrow & \text{det.no.} & 1 & 2 & 3 & 4 & 0 \\ 1 & 1 & \begin{pmatrix} 0 & 0 & 1 & 1 & 1 \\ 1 & 2 & 0 & 0 & 0 & 0 & 1 \\ 2 & 3 & 1 & 0 & 0 & 0 & 0 \\ 2 & 4 & 1 & 0 & 0 & 0 & 0 \\ 0 & 0 & 1 & 1 & 0 & 0 & 0 \end{pmatrix} \end{matrix}$$

The decision input vector I_1^t for DM 1 is obtained from the first column. The column time index is given by $c_1 = 1$ and

$$I_1^t = \{u_3^{t+2-1-1}, u_4^{t+2-1-1}, u_0^{t+0-1-1}\} = \{u_3^t, u_4^t, u_0^{t-2}\}.$$

In the D matrix above, time indices and detector numbers appear both as rows and as columns. For brevity, from now on this information will only be presented as columns. The generalized representation is further illustrated by considering the following examples.

Example 4.7

In this example, we look at the serial network of Example 4.4 and obtain the time indices. The communication matrix D is given by

$$D = \begin{array}{c} \text{time index} \\ \downarrow \\ N-1 \\ N-2 \\ N-3 \\ \vdots \\ 1 \\ 0 \end{array} \quad \begin{array}{c} \text{det.no.} \\ 1 \\ 2 \\ 3 \\ \vdots \\ N-1 \\ N \end{array} \begin{pmatrix} 0 & 1 & 0 & 0 & \cdots & 0 \\ 0 & 0 & 1 & 0 & \cdots & 0 \\ 0 & 0 & 0 & 1 & \cdots & 0 \\ \vdots & \vdots & \cdots & \ddots & \ddots & 0 \\ 0 & 0 & 0 & 0 & \vdots & 1 \\ 0 & \cdots & \cdots & \cdots & \cdots & 0 \end{pmatrix}$$

The time index c_k (time index of the detector corresponding to column k) of the non-zero entries in the matrix D can be written in terms of the time index r_i (time index of the detector corresponding to row i) as follows:

$$c_k = r_i - 1.$$

Hence, the input decision vector I_n^t consists of one decision, namely, the previous detector decision

$$I_n^t = \left\{ u_{n-1}^{t+r_i-c_k-1} \right\} = \left\{ u_{n-1}^t \right\}, \; n=2, \, ..., \, N \, .$$

This indicates that the decision of the $(n-1)$th detector is used by the nth detector without any further delay.

Example 4.8

We consider the parallel fusion network as given in Example 4.5 and obtain the time indices. The communication matrix D is of dimension $(N+1)\times(N+1)$ and is given by

$$D = \begin{matrix} 1 & 1 \\ 1 & 2 \\ \vdots & \vdots \\ 1 & N \\ 0 & 0 \end{matrix} \begin{pmatrix} 0 & 0 & \cdots & 0 & 1 \\ 0 & 0 & \cdots & 0 & 1 \\ \vdots & \vdots & \vdots & \vdots & \vdots \\ 0 & 0 & \cdots & 0 & 1 \\ 0 & 0 & \cdots & 0 & 0 \end{pmatrix}$$

It is seen that all the local detectors have the same time index. The time index r_i of the non-zero entries is given by $r_i=1$, $i=1, 2, ..., N$, and the time index c_0 is given by $c_0 = 0$. Hence, the time parameter of the local decisions at the global decision maker is expressed as

$$t + r_i - c_0 - 1 = t + 1 - 0 - 1 = t .$$

The decision input at the global decision maker I_0^t is, therefore, given by

$$I_0^t = \left\{ u_1^t, u_2^t, ..., u_N^t \right\} .$$

Local detectors have no decision input, as seen before.

Example 4.9

We consider the parallel fusion network with feedback. The system consists of N local detectors and a fusion center as shown in Figure 4.8. Levels associated with detectors in this system are the same as in Example 4.8. Hence, the same time indices are obtained. The communication matrix D is, therefore, given by

4.5 Generalized Formulation for Detection Networks

$$D = \begin{pmatrix} 1 & 1 & 0 & 0 & \cdots & 0 & 1 \\ 1 & 2 & 0 & 0 & \cdots & 0 & 1 \\ \vdots & \vdots & \cdots & \cdots & \cdots & \cdots & \cdots \\ 1 & N & 0 & 0 & \cdots & 0 & 1 \\ 0 & 0 & 1 & 1 & \cdots & 1 & 0 \end{pmatrix}$$

Observe the effect of feedback on the matrix D. The bottom row indicates that there is a communication link from the global decision maker (detector 0) to all the local detectors. Note that the decision input of the local detector corresponding to the column k, $k=1, 2, ..., N$, has a time index of one, i.e., $c_k = 1$. The decision input of the local detector corresponding to the column k is given by

$$I_k^t = \{u_0^{t+r_0-c_k-1}\} = \{u_0^{t+0-1-1}\} = \{u_0^{t-2}\} \text{ for any local detector } k.$$

As seen above, the global decision input to the local detectors has a time parameter of $t-2$ which indicates that two time delays are encountered, the local detector delay and the global decision maker delay. It is important to note that the results in Section 4.4 assumed that the global decision maker does not account for any time delay. Hence, the time parameter of $t-1$ was used for the previous global decision.

The decision input of the global decision maker is obtained using the zeroth column,

$$I_0^t = \{u_1^t, u_2^t, ..., u_N^t\}.$$

The time parameter of the local decisions indicate that all local decisions are used without any time delay.

Using the generalized definition of the communication structure, any decentralized detection system can be represented by a communication matrix. Next, we derive the decision rules of all the detectors in a decentralized detection system whose configuration is specified in terms of its communication matrix.

Derivation of Decision Rules

We consider the binary hypothesis testing problem with the two hypotheses denoted by H_0 and H_1, respectively. Let the number of detectors in the system be $N+1$. The block diagram of any detector, say the kth detector, in a decentralized detection system is shown in Figure 4.20. Due to the dependence of actions on time, we associate a time parameter t with all the system variables. The kth detector of the decentralized detection system operates as follows: At time step t, the kth detector, based on the observation input denoted by y_k^t and the decision input denoted by I_k^t produces the detector decision u_k^t using the decision rule $\gamma_k^t(.)$ as follows:

$$u_k^t = \gamma_k^t(y_k^t, I_k^t) .$$

It should be noted that, in decentralized detection systems with a generalized configuration (cycles included), a detector may operate more than once prior to yielding the final decision. Such a detector may also use its previous observations at times $t-1, t-2, ..., 1$, if the detector has memory. Otherwise the detector uses only the current observation sample at time step t.

We assume that the joint conditional probability density of the observations $p(y_0^t, y_1^t,..., y_N^t | H_j), j = 0, 1; t = 1, 2,..., T$, are known a priori. The problem is to find PBPO decision rules $\gamma_k^t(.)$, $k=0, 1,..., N$, $t=1, 2,..., T$, so as to minimize the cost function $\Re(\Gamma)$ for the final decision u_0^T, where

$$\Gamma = \{\Gamma^t : t=1, 2, ..., T\} ,$$

and Γ^t is defined as

$$\Gamma^t = \{\gamma_k^t(.), \text{ for all } k\} .$$

We again consider the Bayesian formulation where the cost function $\Re(\Gamma)$ is given by

4.5 Generalized Formulation for Detection Networks

Figure 4.20. A detector in a distributed detection network.

$$\mathfrak{R}(\Gamma) = C_{00}P(u_0^T = 0, H_0) + C_{01}P(u_0^T = 0, H_1)$$
$$+ C_{10}P(u_0^T = 1, H_0) + C_{11}P(u_0^T = 1, H_1), \quad (4.5.2)$$

where C_{ij}, $i, j = 0, 1$, is the cost of deciding $u_0^T = H_i$ when the true hypothesis is H_j.

The costs C_{ij}, $i, j = 0, 1$, and the a priori probabilities P_0 and P_1 are assumed to be known. We rewrite Equation (4.5.2) in terms of the system probability of false alarm at time step T, P_F^T, and the system probability of detection at time step T, P_D^T, as follows:

$$\mathfrak{R}(\Gamma) = C_F P(u_0^T = 1 | H_0) - C_D P(u_0^T = 1 | H_1) + C$$
$$= C_F P_F^T - C_D P_D^T + C, \quad (4.5.3)$$

where C_F, C_D and C are as defined earlier and have the same properties. It should be noted that the form of $\mathfrak{R}(\Gamma)$ given in Equation (4.5.3) is independent of the system structure (configuration). We assume that the observations of the distributed detection system are spatially, as well as temporally independent, given any hypothesis. Hence, the a priori knowledge of the conditional probability density functions $p(y_0^t, y_1^t, ..., y_N^t | H_j)$, $j=0, 1$; $t=1, 2, ..., T$, reduces to the a priori knowledge of the individual detector conditional probability densities $p(y_k^t | H_j)$, $j=0, 1$; $t=1, 2, ..., T$; $k=0, 1, ..., N$. Next, we proceed with the minimization of the cost function given in Equation (4.5.3). We derive the PBPO decision rule $\gamma_k^t(.)$ for the kth detector with inputs and outputs as shown in Figure 4.20.

Theorem 4.1

For the binary hypothesis testing problem in a generalized decentralized detection system, the PBPO decision rule of the kth detector (Figure 4.20) at time step t that minimizes the Bayesian cost function associated with the global decision at the final time T is given by

$$\gamma_k^t(y_k^t, I_k^t) = u_k^t = \begin{cases} 1, & \text{if } \Lambda(y_k^t) > \eta_k^t(I_k^t) \\ 0, & \text{otherwise}, \end{cases} \quad (4.5.4)$$

for all $k=0, 1, ..., N$; $t=1, 2, ..., T$; where $\eta_k^t(I_k^t)$ is the threshold of the kth detector at time step t defined as

$$\eta_k^t(I_k^t) = \frac{C_F g\,'(T, 0) f\,'(u_k^t, 0) P(I_k^t|H_0)}{C_D g\,'(T, 1) f\,'(u_k^t, 1) P(I_k^t|H_1)}, \quad (4.5.5)$$

and

$$g\,'(T, i) = P(u_0^T = 1 | u_0^t = 1, H_i) - P(u_0^T = 1 | u_0^t = 0, H_i), \quad i = 0, 1,$$

$$f\,'(u_k^t, i) = P(u_0^t = 1 | u_k^t = 1, H_i) - P(u_0^t = 1 | u_k^t = 0, H_i), \quad i = 0, 1.$$

Proof

We start with Equation (4.5.3) and expand it in terms of the decision of the kth detector at time step t, u_k^t, the decision input I_k^t, the observation y_k^t, and the global decision at time step t, u_0^t, as follows:

$$\Re(\Gamma) = \sum_{I_k^t, u_0^t, u_k^t} \int_{y_k^t} C_F P(u_0^T = 1, I_k^t, u_0^t, u_k^t, y_k^t | H_0)$$

$$- C_D P(u_0^T = 1, I_k^t, u_0^t, u_k^t, y_k^t | H_1) + C.$$

(4.5.6)

4.5 Generalized Formulation for Detection Networks

Conditioning on u_0^t, I_k^t, u_k^t, and y_k^t, Equation (4.5.6) is rewritten as

$$\Re(\Gamma) = \sum_{I_k^t, u_0^t, u_k^t} \int_{y_k^t} C_F P(u_0^T = 1 | u_0^t, u_k^t, I_k^t, y_k^t, H_0)$$

$$\times p(u_0^t, u_k^t, y_k^t, I_k^t | H_0)$$

$$- C_D P(u_0^T = 1 | u_0^t, u_k^t, I_k^t, y_k^t, H_1) \times p(u_0^t, u_k^t, I_k^t, y_k^t | H_1) + C .$$

(4.5.7)

Writing the cost function $\Re(\Gamma)$ of (4.5.7) explicitly in terms of all the possibilities of the global decision u_0^t and conditioning further on u_k^t, I_k^t, and y_k^t,

$$\Re(\Gamma) = \sum_{I_k^t, u_k^t} \int_{y_k^t} C_F \, P(u_0^T = 1 | u_0^t = 1, u_k^t, I_k^t, y_k^t, H_0)$$

$$\times P(u_0^t = 1 | u_k^t, y_k^t, I_k^t, H_0) p(u_k^t, y_k^t, I_k^t | H_0)$$

$$- C_D \, P(u_0^T = 1 | u_0^t = 1, u_k^t, I_k^t, y_k^t, H_1)$$

$$\times P(u_0^t = 1 | u_k^t, I_k^t, y_k^t, H_1) p(u_k^t, I_k^t, y_k^t | H_1)$$

$$+ C_F \, P(u_0^T = 1 | u_0^t = 0, u_k^t, I_k^t, y_k^t, H_0)$$

$$\times P(u_0^t = 0 | u_k^t, I_k^t, y_k^t, H_0) p(u_k^t, I_k^t, y_k^t | H_0)$$

$$- C_D P(u_0^T = 1 | u_0^t = 0, u_k^t, I_k^t, y_k^t, H_1)$$

$$\times P(u_0^t = 0 | u_k^t, I_k^t, y_k^t, H_1) p(u_k^t, I_k^t, y_k^t | H_1) + C .$$

(4.5.8)

We observe that the final global decision $u_0^T = 1$ given the global decision at time t, $u_0^t = j$ and the hypothesis H_i does not depend on u_k^t, I_k^t, and y_k^t. Hence, we rewrite (4.5.8) by factoring out the common terms and substituting $P(u_0^t = 0|.)$ by $1 - P(u_0^t = 1|.)$:

170   4. Distributed Bayesian Detection: Other Network Topologies

$$\Re(\Gamma) = \sum_{I_k^t, u_k^t} \int_{y_k^t} C_F p(u_k^t, y_k^t, I_k^t | H_0) \big[P(u_0^T = 1 | u_0^t = 1, H_0)$$

$$\times P(u_0^t = 1 | u_k^t, y_k^t, I_k^t, H_0) + P(u_0^T = 1 | u_0^t = 0, H_0)$$

$$\times (1 - P(u_0^t = 1 | u_k^t, I_k^t, y_k^t, H_0) \big]$$

$$- C_D p(u_k^t, y_k^t, I_k^t | H_1) \big[P(u_0^T = 1 | u_0^t = 1, H_1)$$

$$\times P(u_0^t = 1 | u_k^t, y_k^t, I_k^t, H_1) + P(u_0^T = 1 | u_0^t = 0, H_1)$$

$$\times (1 - P(u_0^t = 1 | u_k^t, I_k^t, y_k^t, H_1)) \big] + C .$$

(4.5.9)

Multiplying out the term $[1 - P(u_0^t = 1|.)]$ and rearranging,

$$\Re(\Gamma) = \sum_{I_k^t, u_k^t} \int_{y_k^t} C_F \, p(u_k^t, y_k^t, I_k^t | H_0) P(u_0^t = 1 | u_k^t, y_k^t, I_k^t, H_0)$$

$$\times \big[P(u_0^T = 1 | u_0^t = 1, H_0) - P(u_0^T = 1 | u_0^t = 0, H_0) \big]$$

$$+ C_F p(u_k^t, y_k^t, I_k^t | H_0) P(u_0^T = 1 | u_0^t = 0, H_0)$$

$$- C_D p(u_k^t, y_k^t, I_k^t | H_1) P(u_0^t = 1 | u_k^t, y_k^t, I_k^t, H_1)$$

$$\times \big[P(u_0^T = 1 | u_0^t = 1, H_1) - P(u_0^T = 1 | u_0^t = 0, H_1) \big]$$

$$- C_D p(u_k^t, y_k^t, I_k^t | H_1) P(u_0^T = 1 | u_0^t = 0, H_1) + C .$$

(4.5.10)

Letting $P(u_0^T = 1 | u_0^t = 1, H_i) - P(u_0^T = 1 | u_0^t = 0, H_i) = g^t(T, i)$, $i=0, 1$, and conditioning (4.5.10) further on y_k^t and I_k^t,

4.5 Generalized Formulation for Detection Networks

$$\begin{aligned}
\Re(\Gamma) = \sum_{I_k^t, u_k^t} \int_{y_k^t} & C_F P(u_k^t | y_k^t, I_k^t, H_0) p(y_k^t, I_k^t | H_0) \\
& \times P(u_0^t = 1 | u_k^t, y_k^t, I_k^t, H_0) g'(T, 0) \\
& + C_F P(u_k^t | y_k^t, I_k^t, H_0) p(y_k^t, I_k^t | H_0) P(u_0^T = 1 | u_0^t = 0, H_0) \\
& - C_D P(u_k^t | y_k^t, I_k^t, H_1) p(y_k^t, I_k^t | H_1) \\
& \times P(u_0^t = 1 | u_k^t, y_k^t, I_k^t, H_1) g'(T, 1) \\
& - C_D P(u_k^t | y_k^t, I_k^t, H_1) p(y_k^t, I_k^t | H_1) P(u_0^T = 1 | u_0^t = 0, H_1) + C .
\end{aligned}$$

(4.5.11)

We note that the kth detector decision u_k^t given the observation y_k^t and the decision input I_k^t does not depend on the hypothesis present.

Rewriting the cost function $\Re(\Gamma)$ of (4.5.11) in terms of all possibilities of the decision u_k^t,

$$\begin{aligned}
\Re(\Gamma) = \sum_{I_k^t} \int_{y_k^t} & P(u_k^t = 1 | y_k^t, I_k^t) \Big[C_F p(y_k^t, I_k^t | H_0) \\
& \times \{ P(u_0^t = 1 | u_k^t = 1, y_k^t, I_k^t, H_0) g'(T, 0) + P(u_0^T = 1 | u_0^t = 0, H_0) \} \\
& - C_D p(y_k^t, I_k^t | H_1) \times \{ P(u_0^t = 1 | u_k^t = 1, y_k^t, I_k^t, H_1) g'(T, 1) \\
& + P(u_0^T = 1 | u_0^t = 0, H_1) \} \Big] + P(u_k^t = 0 | y_k^t, I_k^t) \Big[C_F p(y_k^t, I_k^t | H_0) \\
& \times \{ P(u_0^t = 1 | u_k^t = 0, y_k^t, I_k^t, H_0) g'(T, 0) + P(u_0^T = 1 | u_0^t = 0, H_0) \} \\
& - C_D p(y_k^t, I_k^t | H_1) \times \{ P(u_0^t = 1 | u_k^t = 0, y_k^t, I_k^t, H_1) g'(T, 1) \\
& + P(u_0^T = 1 | u_0^t = 0, H_1) \} \Big] + C .
\end{aligned}$$

(4.5.12)

We observe that the global decision at time t, u_0^t, given $u_k^t = j$ and the hypothesis H_i, does not depend on y_k^t and I_k^t. In addition, replacing $P(u_k^t$

172 4. Distributed Bayesian Detection: Other Network Topologies

$=0|y_k^t,I_k^t)$ by $1-P(u_k^t=1|y_k^t,I_k^t)$ in (4.5.12) and rearranging,

$$\Re(\Gamma) = \sum_{I_k^t} \int_{y_k^t} P(u_k^t=1|y_k^t, I_k^t) \Big[C_F p(y_k^t, I_k^t|H_0)$$

$$\times \left\{ P(u_0^t=1|u_k^t=1, H_0) g'(T, 0) + P(u_0^T=1|u_0^t=0, H_0) \right\}$$

$$- C_D p(y_k^t, I_k^t|H_1) \left\{ P(u_0^t=1|u_k^t=1, H_1) g'(T, 1) \right.$$

$$\left. + P(u_0^T=1|u_0^t=0, H_1) \right\} - C_F p(y_k^t, I_k^t|H_0)$$

$$\times \left\{ P(u_0^t=1|u_k^t=0, H_0) g'(T, 0) + P(u_0^T=1|u_0^t=0, H_0) \right\}$$

$$+ C_D p(y_k^t, I_k^t|H_1) \left\{ P(u_0^t=1|u_k^t=0, H_1) g'(T, 1) \right.$$

$$\left. + P(u_0^T=1|u_0^t=0, H_1) \right\} \Big] + C_F p(y_k^t, I_k^t|H_0)$$

$$\times \left\{ P(u_0^t=1|u_k^t=0, H_0) g'(T, 0) + P(u_0^T=1|u_0^t=0, H_0) \right\}$$

$$- C_D p(y_k^t, I_k^t|H_1) \times \left\{ P(u_0^t=1|u_k^t=0, H_1) g'(T, 1) \right.$$

$$\left. + P(u_0^T=1|u_0^t=0, H_1) \right\} + C .$$

(4.5.13)

We observe that the last three additive terms of Equation (4.5.13) are fixed so far as the optimization of the kth detector is concerned. We ignore these terms in the subsequent equations and denote the remaining terms of the cost function by $\Re^1(\Gamma)$. Rearranging $\Re^1(\Gamma)$ by further factorization of common terms,

4.5 Generalized Formulation for Detection Networks 173

$$\Re^1(\Gamma) = \sum_{I_k^t} \int_{y_k^t} P(u_k^t=1|y_k^t, I_k^t)\{C_F p(y_k^t, I_k^t|H_0)$$

$$\times \left[P(u_0^t=1|u_k^t=1, H_0)g\,'(T, 0) + P(u_0^T=1|u_0^t=0, H_0)\right.$$

$$\left. - P(u_0^t=1|u_k^t=0, H_0)g\,'(T, 0) - P(u_0^T=1|u_0^t=0, H_0)\right]$$

$$- C_D p(y_k^t, I_k^t|H_1)\left[P(u_0^t=1|u_k^t=1, H_1)g\,'(T, 1)\right.$$

$$+ P(u_0^T=1|u_0^t=0, H_1) - P(u_0^t=1|u_k^t=0, H_1)g\,'(T, 1)$$

$$\left.\left. - P(u_0^T=1|u_0^t=0, H_1)\right]\right\}.$$

(4.5.14)

Canceling out equal terms and rearranging by further factorization of common terms, Equation (4.5.14) is rewritten as

$$\Re^1(\Gamma) = \sum_{I_k^t} \int_{y_k^t} P(u_k^t=1|y_k^t, I_k^t)\{C_F p(y_k^t, I_k^t|H_0)g\,'(T,0)$$

$$\times \left[P(u_0^t=1|u_k^t=1, H_0) - P(u_0^t=1|u_k^t=0, H_0)\right]$$

$$- C_D p(y_k^t, I_k^t|H_1)g\,'(T, 1)$$

$$\times \left[P(u_0^t=1|u_k^t=1, H_1) - P(u_0^t=1|u_k^t=0, H_1)\right]\}.$$

(4.5.15)

Letting $P(u_0^t = 1|u_k^t = 1, H_i) - P(u_0^t = 1|u_k^t = 0, H_i) = f'(u_k^t, i)$, $i=0, 1$, we rewrite (4.5.15) as

$$\Re^1(\Gamma) = \sum_{I_k^t} \int_{y_k^t} P(u_k^t=1|y_k^t, I_k^t)$$

$$\times \left[C_F p(y_k^t, I_k^t|H_0)g\,'(T, 0)f'(u_k^t, 0)\right.$$

$$\left. - C_D p(y_k^t, I_k^t|H_1)g\,'(T, 1)f'(u_k^t, 1)\right]. \quad (4.5.16)$$

The cost function $\Re^1(\Gamma)$ of (4.5.16) is minimized if we choose

$$P(u_k^t = 1 | y_k^t, I_k^t) = \begin{cases} 1, & \text{if } A_1 > A_0, \\ 0, & \text{otherwise}, \end{cases} \quad (4.5.17)$$

where

$$A_1 = C_D p(y_k^t, I_k^t | H_1) g^{\,t}(T, 1) f^{\,t}(u_k^t, 1),$$

and

$$A_0 = C_F p(y_k^t, I_k^t | H_0) g^{\,t}(T, 0) f^{\,t}(u_k^t, 0).$$

The kth detector decision rule $\gamma_k^t(.)$ of the general distributed detection system is given by rewriting (4.5.17) as

$$\gamma_k^t(y_k^t, I_k^t) = u_k^t = \begin{cases} 1, & \text{if } \Lambda(y_k^t, I_k^t) > \mu_k^t, \\ 0, & \text{otherwise}, \end{cases} \quad (4.5.18)$$

where μ_k^t is the threshold of the kth detector at time step t defined as

$$\mu_k^t = \frac{C_F g^{\,t}(T, 0) f^{\,t}(u_k^t, 0)}{C_D g^{\,t}(T, 1) f^{\,t}(u_k^t, 1)}.$$

Using the assumption of temporal and spatial conditional independence of observations in the general distributed detection system, the kth detector observation y_k^t is independent of the kth detector decision input I_k^t. Hence, the likelihood ratio is separable as follows:

$$\Lambda(y_k^t, I_k^t) = \Lambda(y_k^t) \times \Lambda(I_k^t).$$

Substituting this result in Equation (4.5.18) and rearranging,

$$\gamma_k^t(y_k^t, I_k^t) = u_k^t = \begin{cases} 1, & \text{if } \Lambda(y_k^t) > \eta_k^t(I_k^t), \\ 0, & \text{otherwise}, \end{cases} \quad (4.5.19)$$

where $\eta_k^t(I_k^t)$ is a multivalued threshold of the kth detector at time step t defined as

$$\eta_k^t(I_k^t) = \frac{C_F g^{\prime t}(T, 0) f^{\prime t}(u_k^t, 0) P(I_k^t | H_0)}{C_D g^{\prime t}(T, 1) f^{\prime t}(u_k^t, 1) P(I_k^t | H_1)}, \quad (4.5.20)$$

as given in Equation (4.5.5).

It should be noted that Equation (4.5.18) represents the decision rule for any detector of a decentralized detection system configured in any manner. Moreover, the decision rule of any detector k at any time step t is based on the likelihood ratio of the input to that detector, and this includes the decision rule at the global decision maker. The result for the global decision maker is stated next as a corollary.

Corollary 4.1

For a general decentralized detection system, the PBPO decision rule $\gamma_0^T(.)$ of the global decision maker that minimizes the Bayesian cost function for the binary hypothesis testing problem is given by

$$\gamma_0^T(I_0^T, y_0^T) = u_0^t = \begin{cases} 1, & \text{if } \Lambda(I_0^T, y_0^T) > \dfrac{C_F}{C_D}, \\ 0, & \text{otherwise}, \end{cases} \quad (4.5.21)$$

where I_0^T is the decision input of the global decision maker and y_0^T is the input observation of the global decision maker (if any).

Proof

The global decision rule of (4.5.21) results directly from the general decision rule (4.5.18) by letting $k = 0$, $t = T$, and observing the following

$$g^T(T, i) = P(u_0^T = 1 | u_0^T = 1, H_j) - P(u_0^T = 1 | u_0^T = 0, H_j) ,$$
$$= 1 - 0 = 1$$

and, for $t = 1, 2, \ldots, T$,

$$f'(u_0^t, i) = P(u_0^t = 1 | u_0^t = 1, H_j) - P(u_0^t = 1 | u_0^t = 0, H_j)$$
$$= 1.$$

Hence, the threshold of (4.5.18) reduces to

$$\eta_k^T = \frac{C_F}{C_D} ,$$

resulting in Equation (4.5.21).

Note that the decision rule of (4.5.21) is a general global decision rule in that the global decision maker may also make direct observations of the phenomenon in addition to the decisions received from the other detectors. The observation term y_0^T is to be deleted if there is no direct observation at the global decision maker.

The result of Theorem 4.1 can be applied to a variety of detection networks involving memory and feedback. As a further example, we consider the tree configuration discussed in Section 4.3. In this configuration, detector decisions flow in one direction only, namely, toward the global decision maker. Hence, a detector operates only once and the time parameter t need not be taken into account. The result is presented in Corollary 4.2.

Corollary 4.2

For the binary hypothesis testing problem for a detection network in a tree configuration, the PBPO decision rule at the kth detector (Figure 4.20) that minimizes the Bayesian cost function of the final global decision is given by

$$\gamma_k(y_k, I_k) = \begin{cases} 1, & \text{if } \Lambda(y_k) > \eta_k(I_k), \\ 0, & \text{otherwise}, \end{cases} \quad (4.5.22)$$

where I_k is the decision input of the kth detector and $\eta_k(I_k)$ is the threshold of the kth detector defined as

$$\eta_k(I_k) = \frac{C_F f(u_k, 0) P(I_k|H_0)}{C_D f(u_k, 1) P(I_k|H_1)}, \quad (4.5.23)$$

and

$$f(u_k, i) = P(u_0 = 1 | u_k = 1, H_i) - P(u_0 = 1 | u_k = 0, H_i), i = 0, 1.$$

Proof

This result is obtained simply by dropping the superscript t in the result of Theorem 4.1, because in this formulation, a detector operating at two different time instants is considered as two different detectors. The term $g^t(T, i) = 1$ because the global decision maker operates only once, namely at $t=T$.

It should be noted that the result of Corollary 4.2 agrees with that of Reibman and Nolte [ReN87b]. The generalized formulation presented in this section has been employed to obtain decision rules for a variety of distributed detection networks as special cases in [Alh90]. Network configurations considered include fairly complex structures, such as a network with peer communication, where all the local detectors are able to communicate with each other. In this configuration, observations arrive sequentially at the local detectors. As observations arrive, local detectors make their decisions and transmit them to the fusion center as well as to all other detectors. Decision at a local detector is based on its present observation, its past observations, and the most recent set of decisions received from the other detectors. Several other network structures that involve communication amongst all detectors have been analyzed in the literature. For example, in the structure considered in [ChV88], each local detector makes a tentative decision based on all of its observations. This tentative decision is transmitted to all other detectors. Based on its original observation and the set of decisions received, each local detector makes its final decision. A generalized version of this structure has been analyzed in [SwW95] where multiple transmissions of tentative decisions or "parleying" may take place. In this case, each local detector transmits its tentative decision to all other detectors. Based on its original observation and on the most recent set of tentative decisions received, each local detector "rethinks" and makes another tentative decision. This process continues until all detectors agree and reach a consensus. In this system, there are two considerations, namely, the time it takes to reach a consensus and the correctness of the final result. System design, convergence issues, and an interesting study of the trade-offs involved are presented in [SwW95]. Other distributed detection network configurations can be conceived and analyzed based on the methodology presented in this section.

Notes and Suggested Reading

Design and performance of serial detection networks are discussed in [Rei 87, Tan90, Pap90, TPK91b, Swa93]. Design of tree networks is treated in [Rei87, TPK93, Tsi93a]. Bayesian formulation of the parallel fusion network with feedback was analyzed in [Alh90, AlV90, AlV96]. Representation methodology for distributed detection network topological structures and their design are described in [Alh90, AlV95].

5
Distributed Detection with False Alarm Rate Constraints

5.1 Introduction

In this chapter, we consider the distributed detection problem for situations where the probability of false alarm is to remain less than an acceptable value. This formulation is especially suitable for radar applications. First, we consider the Neyman–Pearson formulation of the problem. This formulation does not require the knowledge of a priori probabilities associated with different hypotheses or an assignment of costs to different courses of action. System probability of detection is maximized under a probability of false alarm constraint. Under this formulation, the parallel fusion network topology without a fusion center is not appropriate because systemwide probabilities of detection and false alarm can not be defined. Therefore, a fusion center is always assumed to be present. We consider only the parallel fusion network topology here. Other network topologies can be treated similarly. In Section 5.2, the distributed Neyman–Pearson detection problem is formulated and decision rules are derived. A number of interesting issues, such as randomization, arise. These and other related aspects are discussed. In practical radar signal detection scenarios, noise and clutter background are often nonstationary. In this case, the optimum Neyman–Pearson detector with a fixed threshold fails to maintain a constant false alarm rate (CFAR), and adaptive thresholding based on observations from the neighboring region is required. Distributed CFAR processing is discussed in Section 5.3. Issues, such as robustness in the presence of homogeneous and nonhomogeneous backgrounds, are also examined. In Section 5.4, distributed detection of weak signals is considered, and locally optimum decision rules are derived.

5.2 Distributed Neyman–Pearson Detection

In conventional, centralized, hypothesis testing problems, it is always possible to design equivalent Bayesian and Neyman–Pearson tests. For a Neyman–Pearson test designed for a specified P_F, an equivalent Bayesian test with appropriate cost assignment and prior probabilities can always be found. But in decentralized detection problems, the situation is a bit more involved. It may not always be possible to design equivalent decentralized Bayesian and decentralized Neyman–Pearson tests. This is because the desired value of P_F at the fusion center may not be achievable by an optimum, decentralized, nonrandomized Bayesian strategy. It is, however, possible to attain the desired value of P_F by randomization of Bayesian strategies and by incorporating additional communication/coordination among the sensors and the fusion center. These tests are known as tests with dependent randomization [Tsi93a] or scheduled tests [WiW92]. Their implementation requires synchronization among detectors, and any loss of synchronization results in performance degradation. Tests with dependent randomization will not be considered in detail in this book.

We consider the Neyman–Pearson formulation of the binary hypothesis testing problem for the parallel fusion network topology shown in Figure 3.6. The same notations and system structure as in Section 3.4 are employed. Based on conditionally independent observations, local detectors transmit their hard decisions to the fusion center that makes the global decision. The observations are assumed to have a continuous distribution, so that the likelihood ratios do not contain any point masses. This assumption results in some simplifications with regard to randomization at the fusion center. The situation, when this assumption does not hold, is discussed in [Tsi93a]. The objective, here, is to design the fusion rule and the local decision rules that maximize the power of the system, i.e., the probability of detection P_D at the fusion center (or, equivalently minimize the probability of miss P_M), while keeping the probability of false alarm P_F at a desired level, say α. One may consider the design of the overall system, i.e., the design of both the fusion rule and the local decision rules, simultaneously, or one may consider subproblems, such as the design of local decision rules, given a specific fusion rule.

First, we determine the structure of the decision rules. We consider only the case in which hard decisions (binary messages) are sent from the local detectors to the fusion center. The fusion center receives the set of local decisions $u_1, ..., u_N$ and makes the final decision u_0, so that the power of the system is maximum for a given level. From the Neyman–Pearson lemma, the fusion rule is the standard Neyman–Pearson test of the form

$$u_0 = \begin{cases} 1, & \text{if } P(u|H_1) > kP(u|H_0), \\ 0, & \text{if } P(u|H_1) < kP(u|H_0) \, . \end{cases} \quad (5.2.1)$$

Note that we have considered only deterministic fusion rules. Next, we show that the decision rules at the local detectors are also Neyman–Pearson tests based on likelihood ratios. Recall from Remark 3.3.1 that the optimum fusion rule is a monotonic function. Let us assume that the tests at the local detectors are operating so that the overall probability of false alarm P_F is equal to the desired level α but the tests are not Neyman–Pearson tests. These tests yield the probabilities of false alarm and detection P_{Fi} and P_{Di}, $i = 1, ..., N$, respectively. As noted in Remark 3.3.2, P_F and P_D depend only on P_{Fi} and P_{Di}, $i=1, ..., N$, respectively. The constraint $P_F = \alpha$ induces constraints on individual P_{Fi} and determines levels at each of the local detectors. Recall from Remark 3.3.2 that P_D is an increasing function of the P_{Di}, $i=1, ..., N$. Therefore, to obtain the maximum value of P_D, each local detector should operate at its maximum achievable P_{Di}, i.e., use the most powerful test, for its given level P_{Fi}. Otherwise, the global P_D can be increased by using a test at the local detector with a higher P_{Di}. Thus, the tests at the local detectors are most powerful for given levels P_{Fi}, $i=1, ..., N$, i.e., they are Neyman–Pearson tests of the form

$$u_i = \begin{cases} 1, & \text{if } p(y_i|H_1) > t_i p(y_i|H_0) \, , \\ 0, & \text{if } p(y_i|H_1) < t_i p(y_i|H_0) \, . \end{cases} \quad (5.2.2)$$

Therefore, it suffices to determine thresholds for LRTs at the fusion center and at the local detectors.

Remark 5.2.1

We have considered only the case where binary messages are transmitted from local detectors to the fusion center. The more general case, in which soft decisions are transmitted, has been considered in [Tsi93b, WaW89, VAT88]. It has been shown that decisions at local detectors are obtained by likelihood ratio partitioning.

Remark 5.2.2

We have limited our attention to deterministic fusion rules. In centralized Neyman–Pearson detection, randomization is employed to be able to attain any specified value of P_F and to make the ROC concave. Recall the discussion in Remark 3.4.1. The ROCs corresponding to the AND and the OR fusion rules were concave, but the team ROC, defined as the maximum of the two individual ROCs, was not concave. To make the team ROC concave, one can consider randomization along the tangent OA (see Figure 3.13). But this randomization is complicated because it involves two different fusion rules and, therefore, different sets of local thresholds. Clearly, this type of randomization requires synchronization between local detectors and was earlier referred to as dependent randomization. It is certainly feasible to implement such tests but they are not preferred due to the additional complexity. In addition, such tests may not perform well if synchronization is lost. Willett and Warren [WiW92] investigated the need for randomization at the fusion center and have shown that, under the assumption that local likelihood ratios contain no point masses of probability, it is not necessary to consider randomized fusion rules. Their result shows that if a randomized fusion rule is required to achieve a desired false alarm rate for given decision rules, the same value of P_F can be achieved by employing a deterministic fusion rule and modifying the local decision rules appropriately. This results in considerable simplification of system design. If, however, local likelihood ratios contain point masses of probability, randomized fusion rules and randomized local decision rules need to be considered.

Derivation of Decision Rules

As seen earlier, the decision rules that maximize P_D for a given value of P_F are likelihood ratio tests with thresholds to be determined. An approach to solving this type of constrained optimization problem is to employ the method of Lagrange multipliers. To minimize a function $f(x)$ subject to a constraint $g(x) = 0$, the Lagrangian $F(x) = f(x) + \lambda g(x)$ is minimized without constraints. If $F(x)$ is convex, this method yields the minimum. Otherwise, the optimization procedure may fail. In this case, Hestenes [Hes75] has suggested convexification of the Lagrangian $F(x)$ and the use of augmented function $H(x) = f(x) + \lambda g(x) + (\sigma/2)\, g^2(x)$. Here, we employ the Lagrangian to derive the decision rules. Convexification of the Lagrangian or other constrained optimization methods may be used in the event that this method fails. Local likelihood ratios are assumed to contain no point masses of probability, and only deterministic decision rules are considered.

The probabilities P_F, P_M, and P_D can be expressed in terms of P_{Fi} and P_{Mi} by defining

$$M_u = P(u|H_1) = \prod_{S_0} P_{Mj} \prod_{S_1} (1 - P_{Mk}), \qquad (5.2.3)$$

$$F_u = P(u|H_0) = \prod_{S_0} (1 - P_{Fj}) \prod_{S_1} P_{Fk}, \qquad (5.2.4)$$

and

$$P_{iu} = P(u_0 = i | u), \quad i = 0, 1, \qquad (5.2.5)$$

where S_0 and S_1 are as defined in Section 3.3. Then, we may express P_M and P_F as follows:

$$P_M = \sum_u P_{0u} M_u, \qquad (5.2.6)$$

and

$$P_F = \sum_u P_{1u} F_u. \qquad (5.2.7)$$

We employ the Lagrange multiplier method for this constrained optimization problem by constructing the objective function F as follows:

$$F = P_M + \Gamma(P_F - \alpha), \qquad (5.2.8)$$

where Γ is the Lagrange multiplier. We express

$$\begin{aligned}
F &= \sum_u P_{0u} M_u + \Gamma (\sum_u P_{1u} F_u - \alpha) \\
&= \sum_u P_{0u} M_u + \Gamma (\sum_u (1-P_{0u}) F_u - \alpha) \\
&= \Gamma(1-\alpha) + \sum_u P_{0u}[P(u|H_1) - \Gamma P(u|H_0)]. \quad (5.2.9)
\end{aligned}$$

F is minimized if we choose the decision rule

$$P_{0u} = P(u_0 = 0|u) = \begin{cases} 0, & \text{if } P(u|H_1) - \Gamma P(u|H_0) > 0, \\ 1, & \text{otherwise}. \end{cases} \qquad (5.2.10)$$

This may be rewritten as

$$\frac{P(u|H_1)}{P(u|H_0)} \underset{u_0=0}{\overset{u_0=1}{\gtrless}} \Gamma. \qquad (5.2.11)$$

Due to the independence assumption,

$$\prod_{j=1}^{N} \frac{P(u_j|H_1)}{P(u_j|H_0)} \overset{u_0=1}{\underset{u_0=0}{\overset{>}{<}}} \Gamma. \quad (5.2.12)$$

By definition,

$$P(u_j|H_1) = (1-P_{Mj})^{u_j} P_{Mj}^{1-u_j}, \quad (5.2.13)$$

and

$$P(u_j|H_0) = P_{Fj}^{u_j} (1-P_{Fj})^{1-u_j}. \quad (5.2.14)$$

Substituting these in (5.2.12),

$$\prod_{j=1}^{N} \frac{(1-P_{Mj})^{u_j} P_{Mj}^{1-u_j}}{P_{Fj}^{u_j}(1-P_{Fj})^{1-u_j}} \overset{u_0=1}{\underset{u_0=0}{\overset{>}{<}}} \Gamma. \quad (5.2.15)$$

Taking the logarithm,

$$\sum_{j=1}^{N} u_j \log(1-P_{Mj}) + (1-u_j) \log P_{Mj} - u_j \log P_{Fj}$$

$$-(1-u_j) \log(1-P_{Fj}) \overset{u_0=1}{\underset{u_0=0}{\overset{>}{<}}} \log \Gamma. \quad (5.2.16)$$

In a more compact form, it can be expressed as

$$\sum_{j=1}^{N} \left[u_j \log \frac{1-P_{Mj}}{P_{Fj}} + (1-u_j) \log \frac{P_{Mj}}{1-P_{Fj}} \right] \overset{u_0=1}{\underset{u_0=0}{\overset{>}{<}}} \log \Gamma. \quad (5.2.17)$$

Thus, the fusion rule has the same structure as obtained in Chapter 3. A weighted sum of the incoming local decisions (same weights as the Bayesian case) is formed and compared to a threshold log Γ.

Next, we determine the decision rule at the local detector k by employing the person-by-person optimization procedure. We may express $P(u_k|H_i)$ as

$$P(u_k|H_i) = \int_{y_k} P(u_k|y_k, H_i) p(y_k|H_i) dy_k, \quad i=0, 1 . \quad (5.2.18)$$

Because u_k does not depend on which hypothesis is present and depends only on y_k,

$$P(u_k|H_i) = \int_{y_k} P(u_k|y_k) p(y_k|H_i) dy_k, \quad i=0, 1 . \quad (5.2.19)$$

The objective function F can be expanded by explicitly summing over u_k,

$$\begin{aligned} F &= \Gamma(1-\alpha) + \sum_{u} P(u_0=0|u)[P(u|H_1) - \Gamma P(u|H_0)] \\ &= \Gamma(1-\alpha) + \sum_{u^k} P(u_0=0|u_k=0, u^k)[P(u_k=0, u^k|H_1) \\ &\quad -\Gamma P(u_k=0, u^k|H_0)] + P(u_0=0|u_k=1, u^k)[P(u_k=1, u^k|H_1) \\ &\quad -\Gamma P(u_k=1, u^k|H_0)] , \end{aligned} \quad (5.2.20)$$

where u^k is as defined in Chapter 3. Setting

$$P(u_k=1, u^k|H_j) = P(u^k|H_j) - P(u_k=0, u^k|H_j), \quad j=0, 1 ,$$

in (5.2.20),

5.2 Distributed Neyman–Pearson Detection 187

$$F = \Gamma(1-\alpha) + \sum_{u^k} P(u_0=0|u_k=1, u^k)[P(u^k|H_1) - \Gamma P(u^k|H_0)]$$
$$+ \sum_{u^k} [P(u_0=0|u_k=0, u^k) - P(u_0=0|u_k=1, u^k)]P(u_k=0, u^k|H_1)$$
$$- \Gamma[P(u_0=0|u_k=1, u^k) - P(u_0=0|u_k=0, u^k)]P(u_k=0, u^k|H_0) .$$

(5.2.21)

Let

$$C^k = \Gamma(1-\alpha) + \sum_{u^k} P(u_0=0|u_k=1, u^k)[P(u^k|H_1) - \Gamma P(u^k|H_0)] .$$

(5.2.22)

Also note that

$$P(u_k=0, u^k|H_j) = P(u_k=0|u^k, H_j)P(u^k|H_j), \quad j=0,1 .$$

(5.2.23)

Furthermore, u_k does not depend on u^k and, therefore,

$$P(u_k=0|u^k, H_j) = P(u_k=0|H_j) = \int_{y_k} P(u_k=0|y_k)p(y_k|H_j)dy_k .$$

(5.2.24)

Thus,

$$F = C^k + \sum_{u^k} [P(u_0=0|u_k=0, u^k) - P(u_0=0|u_k=1, u^k)]P(u^k|H_1)$$
$$\times \int_{y_k} P(u_k=0|y_k)p(y_k|H_1)dy_k - \Gamma[P(u_0=0|u_k=1, u^k)$$
$$- P(u_0=0|u_k=0, u^k)]P(u^k|H_0) \int_{y_k} P(u_k=0|y_k)p(y_k|H_0)dy_k$$
$$= C^k + \int_{y_k} P(u_k=0|y_k)[C_1^k p(y_k|H_1) - \Gamma C_0^k p(y_k|H_0)]dy_k ,$$

(5.2.25)

where

$$C_i^k = \sum_{u^k} [P(u_0=0|u_k=0, u^k) - P(u_0=0|u_k=1, u^k)] P(u^k|H_i),$$

$$i = 0, 1.$$

Because C^k is independent of the decision rule at the kth detector, F is minimized when we set

$$P(u_k=0|y_k) = \begin{cases} 0, & \text{if } C_1^k p(y_k|H_1) - \Gamma C_0^k p(y_k|H_0) > 0, \\ 1, & \text{otherwise}, \end{cases} \quad (5.2.26)$$

or

$$\Lambda(y_k) = \frac{p(y_k|H_1)}{p(y_k|H_0)} \underset{u_k=0}{\overset{u_k=1}{\gtrless}} t_k, \quad (5.2.27)$$

where

$$t_k = \Gamma \frac{C_0^k}{C_1^k}$$

$$= \Gamma \frac{\sum_{u^k} [P(u_0=0|u_k=0, u^k) - P(u_0=0|u_k=1, u^k)] \prod_{\substack{j=1 \\ j \neq k}}^{N} P_{Fj}^{u_j}(1-P_{Fj})^{1-u_j}}{\sum_{u^k} [P(u_0=0|u_k=0, u^k) - P(u_0=0|u_k=1, u^k)] \prod_{\substack{j=1 \\ j \neq k}}^{N} P_{Dj}^{u_j}(1-P_{Dj})^{1-u_j}}.$$

(5.2.28)

The set of equations (5.2.17) and (5.2.28) may be solved simultaneously under the constraint $P_F = \alpha$ to yield the PBPO solution. In some situations, the above Lagrangian formulation may fail to yield a solution. For example, if the minimum lies on the boundary of the solution space for which one or more of the P_{Fi} are either zero or one, the Lagrange multiplier method need not provide a solution [TVB89], and other approaches for system optimization need to be followed. Frequently, attention is limited to a small subset of nonrandomized fusion rules, such as the AND and the OR rules. In this case, local detector thresholds along with system performance can be evaluated for each fusion rule, and optimum system determined.

Example 5.1

Consider a parallel fusion network consisting of two local detectors with conditionally independent observations. We employ the Lagrange multiplier method for system optimization to illustrate the methodology and to point out its limitations. We consider the AND and the OR fusion rules. For the AND rule, the thresholds and the performance in terms of P_M and P_F are given by

$$t_1 = \Gamma \frac{P_{F2}}{1 - P_{M2}}, \qquad (5.2.29)$$

$$t_2 = \Gamma \frac{P_{F1}}{1 - P_{M1}}, \qquad (5.2.30)$$

$$P_M = P_{M1} + P_{M2} - P_{M1} P_{M2}, \qquad (5.2.31)$$

and

$$P_F = P_{F1} P_{F2}. \qquad (5.2.32)$$

The corresponding relationships for the OR fusion rule are

$$t_1 = \Gamma \frac{1-P_{F2}}{P_{M2}}, \tag{5.2.33}$$

$$t_2 = \Gamma \frac{1-P_{F1}}{P_{M1}}, \tag{5.2.34}$$

$$P_M = P_{M1} P_{M2}, \tag{5.2.35}$$

and

$$P_F = P_{F1} + P_{F2} - P_{F1} P_{F2}. \tag{5.2.36}$$

Note that the fusion rule specifies the relationship between global and local probabilities of false alarm and detection. The Lagrange multiplier Γ appears in the threshold expressions and determines the relationship between the local detector parameters. If a feasible solution exists that is interior to the solution space, then, there must be a value of Γ that satisfies the following relationships. For the AND rule,

$$\Gamma = \frac{(1-P_{M2})t_1}{P_{F2}} = \frac{(1-P_{M1})t_2}{P_{F1}}, \tag{5.2.37}$$

and, for the OR rule,

$$\Gamma = \frac{P_{M2} t_1}{1-P_{F2}} = \frac{P_{M1} t_2}{1-P_{F1}}. \tag{5.2.38}$$

If there is no feasible solution or the solution lies on the boundary, (5.2.37) and (5.2.38) are not satisfied.

Let us consider a slow Rayleigh fading environment. In this case,

$$P_{Fi} = (t_i(1+\varepsilon_i))^{-1-(1/\varepsilon_i)},$$

and

$$P_{Mi} = 1 - P_{Fi}^{1/(1+\varepsilon_i)}, \quad i=1, 2,$$

where ε_i is the average received energy-to-noise ratio at DM i. Let $\alpha =$

10^{-6}. Given the values of ε_1 and ε_2, the thresholds and the system performance are obtained numerically and are shown in Figures 5.1 and 5.2. In Figure 5.1, P_D versus ε_2 with ε_1 as a parameter is shown for the AND and OR fusion rules. For the OR rule, $\varepsilon_1 = 8$ dB, 12 dB, 21 dB, 30 dB, and $\varepsilon_1 = \varepsilon_2$ cases are plotted. For the AND rule, this method yields the solution only for the $\varepsilon_1 = \varepsilon_2$ case and is shown in Figure 5.1. For the $\varepsilon_1 \neq \varepsilon_2$ case with the AND rule, no value of Γ can be found that satisfies (5.2.37). In any case, the OR rule performs better and is to be selected over the AND rule.

Next, consider the case in which the conditional densities at the two detectors are Gaussian. Under H_0, the conditional densities at both detectors are assumed to be identical with mean zero and variance one. Under H_1, the mean at DM j, $j=1, 2$, is given by m_j and the variance at both detectors is unity. Let $\alpha = 0.05$. In Figure 5.3, the global probability of detection as a function of m_2, for the AND and the OR rules, is presented. Three cases are considered, namely, $m_1 = m_2$, $m_1 = 1.8$, and $m_1 = 2.3435$. In all three cases, performance curves for the two fusion rules intersect each other, i.e., for some values of m_1 and m_2, one fusion rule is better than the other, and vice versa. For a given value of m_2, there are ranges of values of m_1 for which feasible solutions exist. For example, for $m_2 = 0.5$, when the OR rule is used, a feasible solution exists only for $0.1179 < m_1 < 2.3435$ and, when the AND rule is employed, a feasible solution exists only for $0.3021 < m_1 < 3.02$. Threshold values at the two sensors for the two fusion rules are shown in Figures 5.4 and 5.5.

Difficulties associated with the computation of decision rules for the decentralized Neyman–Pearson detection problem arise due to the fact that an additional constraint, $P_F = \alpha$, is to be satisfied. In situations where the Lagrange multiplier method fails, one needs to employ other multidimensional optimization procedures for system design as discussed earlier.

5.3 Distributed CFAR Detection

When signal detection is to be performed in a nonstationary clutter and noise background, CFAR processing, based on an adaptive threshold,

5. Distributed Detection with False Alarm Rate Constraints

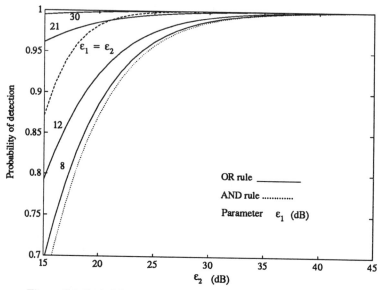

Figure 5.1. Probability of detection as a function of ε_2 for $\alpha = 10^{-6}$.

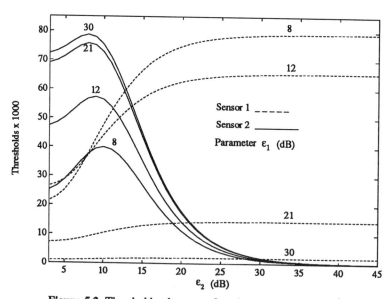

Figure 5.2. Threshold values as a function of ε_2 for $\alpha = 10^{-6}$.

5.2 Distributed Neyman–Pearson Detection

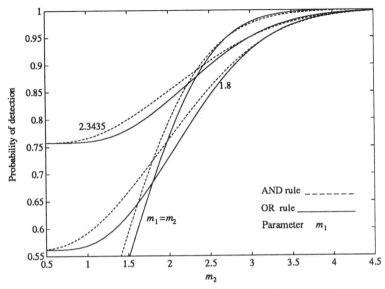

Figure 5.3. Probability of detection as a function of m_2 for $\alpha = 0.05$.

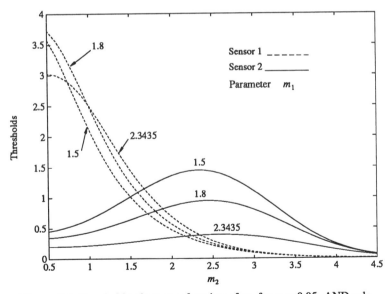

Figure 5.4. Threshold values as a function of m_2 for $\alpha = 0.05$, AND rule.

5. Distributed Detection with False Alarm Rate Constraints

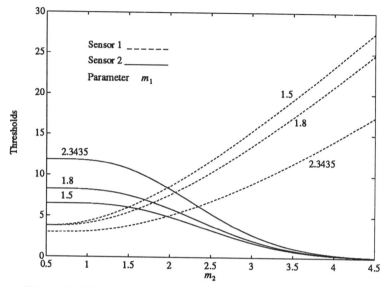

Figure 5.5. Threshold values as a function of m_2 for $\alpha=0.05$, OR rule.

instead of Neyman–Pearson detection, based on a fixed threshold, is carried out. In this section, we address the problem of CFAR processing in a distributed framework. We consider the distributed detection system consisting of N distributed CFAR detectors with a fusion center as shown in Figure 5.6. The notation, terminology, and results on CFAR detection given in Section 2.6 are used here. It is assumed that the reference window size at the ith detector is N_i, $i = 1, 2, ..., N$. The target is assumed to be a slowly fluctuating target of Swerling type I, and the homogeneous background noise is Gaussian with an unknown level. The probability of false alarm and the probability of detection at the individual detectors are denoted by P_{Fi} and P_{Di}, $i = 1, 2, ..., N$, respectively. When there is a target in the test cell, it is assumed that the target SNR at the local detectors is the same, i.e., $S_1 = S_2 = \cdots = S_N = S$. Generalization for the case of unequal local target SNRs can be obtained in a similar manner. The local false alarm probability, P_{Fi}, for detector i, $i = 1, 2, ..., N$, is given by

$$P_{Fi} = \int_0^\infty P[y_i > T_i Z_i | Z_i, H_0] p(Z_i) dZ_i , \qquad (5.3.1)$$

where y_i is the test cell sample, T_i is the scaling factor and $p(Z_i)$ denotes

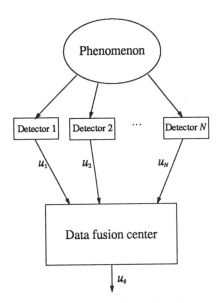

Figure 5.6. A distributed CFAR detection system.

the pdf of Z_i, the estimate of noise power at the local CFAR detector i, $i = 1, 2, ..., N$. Similarly, the detection probability of local detector i is given by

$$P_{Di} = \int_0^\infty P[y_i > T_i Z_i | Z_i, H_1] p(Z_i) dZ_i . \qquad (5.3.2)$$

Each local CFAR detector transmits its decision u_i, $i = 1, ..., N$, to the data fusion center. The global decision u_0 is made, based on the local decisions. The global probabilities of false alarm and miss, i.e., P_F and P_M, are as given in (5.2.6) and (5.2.7). The probabilities P_{0u} and P_{1u} are determined by the fusion rule and the probabilities M_u and F_u are based on the target-noise models and the parameters of the local detectors. Here, we shall consider system design for fixed fusion rules. The goal will be to maximize P_D while maintaining a desired level of P_F at the fusion center. For the target and noise models under consideration, likelihood ratios at the local detectors do not contain point masses and randomized fusion rules need not be considered. We employ the Lagrange multiplier method for this constrained optimization problem. We consider cases where CA-CFAR and OS-CFAR are employed as local detectors along with fusion rules AND and OR. We confine our attention to the case where identical CFAR processing techniques are used at all local detectors. Distributed CFAR detection systems with local detectors employing nonidentical CFAR processing techniques can be designed similarly as discussed in [EBA 92].

CA-CFAR Processors

In the distributed detection system with CA-CFAR local detectors operating in a homogeneous background, the global detection probability, for a given data fusion rule, is maximized by setting the scale factors, T_i, $i = 1, 2, ... N$, of the local detectors optimally. For a given SNR common to all local detectors and a given set of reference window sizes, the Lagrangian, formed as in [BaV89], is

$$J(T_1, T_2, ..., T_N) = P_D(T_1, T_2, ..., T_N) + \Gamma[P_F(T_1, T_2, ..., T_N) - \alpha] ,$$

(5.3.3)

where J is the objective function to be maximized, Γ is the Lagrange multiplier, and α is the desired constant false alarm rate at the fusion center. To determine the probabilities P_D and P_F, the values of P_{Fi} and P_{Di} of the local detectors are required. They are obtained from (2.6.16) and (2.6.18) by replacing N by N_i and T by T_i.

In the optimization procedure, we set the derivatives of $J(T_1, T_2, ..., T_N)$ with respect to T_i, $i = 1, 2, ..., N$, equal to zero, i.e.,

$$\frac{\partial J(T_1, T_2, ..., T_N)}{\partial T_i} = 0, \quad i = 1, 2, ..., N. \qquad (5.3.4)$$

This set of equations and the constraint

$$P_F(T_1, T_2, ..., T_N) = \alpha \qquad (5.3.5)$$

yield $(N + 1)$ nonlinear equations with $(N + 1)$ unknowns where the unknowns are the scaling factors of the local detectors, T_i, $i = 1, 2, ..., N$, and the Lagrange multiplier Γ. The optimum set of T_i, $i = 1, 2, ..., N$, which gives the highest global probability of detection P_D is found by solving equations (5.3.4) and (5.3.5) simultaneously. Once again, recall that this approach yields a solution only if a feasible solution exists interior to the solution space. Otherwise, alternate optimization algorithms need to be used. Now we give specific results for the AND and the OR fusion rules based on the Lagrangian approach.

AND Fusion Rule

For the AND fusion rule,

$$P_D = \prod_{i=1}^{N} P_{Di} , \qquad (5.3.6)$$

and

$$P_F = \prod_{i=1}^{N} P_{Fi} . \qquad (5.3.7)$$

Substituting the expressions for P_{Fi} and P_{Di}, the objective function is given by

$$J(T_1, T_2, ..., T_N) = \prod_{i=1}^{N} \frac{(1+S)^{N_i}}{(1+S+T_i)^{N_i}} + \Gamma [\prod_{i=1}^{N} \frac{1}{(1+T_i)^{N_i}} - \alpha] .$$

$$(5.3.8)$$

Taking the derivative of $J(T_1, T_2, ..., T_N)$ with respect to T_j, $j = 1, 2, ..., N$, and setting it equal to zero,

$$\frac{(1+S)^{N_j}}{(1+S+T_j)^{N_j+1}} \prod_{\substack{i=1 \\ i \neq j}}^{N} \frac{(1+S)^{N_i}}{(1+S+T_i)^{N_i}} + \frac{\Gamma}{(1+T_j)^{N_j+1}} \prod_{\substack{i=1 \\ i \neq j}}^{N} \frac{1}{(1+T_i)^{N_i}}$$

$$= 0, \quad j=1, 2, ..., N. \quad (5.3.9)$$

The threshold multipliers or scale factors, T_i, $i=1, ..., N$, can be obtained by solving the above set of coupled nonlinear equations along with the constraint

$$P_F = \prod_{i=1}^{N} \frac{1}{(1+T_i)^{N_i}} = \alpha. \quad (5.3.10)$$

In the special case when $N = 2$, the solution for the above set of equations is given by [BaV89]

$$T_1 = T_2 = -1 + \alpha^{-\frac{1}{N_1+N_2}}. \quad (5.3.11)$$

OR Fusion Rule

In this case,

$$P_M = \prod_{i=1}^{N} P_{Mi}, \quad (5.3.12)$$

and

$$P_F = 1 - \prod_{i=1}^{N} (1 - P_{Fi}). \quad (5.3.13)$$

The objective function can, then, be expressed as

$$J(T_1, T_2, ..., T_N) = 1 - \prod_{i=1}^{N} \left(1 - \frac{(1+S)^{N_i}}{(1+S+T_i)^{N_i}}\right)$$

$$+ \Gamma\left[1 - \prod_{i=1}^{N}\left(1 - \frac{1}{(1+T_i)^{N_i}}\right) - \alpha\right]. \quad (5.3.14)$$

Taking the derivative of $J(T_1, T_2, ..., T_N)$ with respect to T_j, $j = 1, 2, ..., N$, and setting it equal to zero, we get the set of equations

$$\frac{(1+S)^{N_j}}{(1+S+T_j)^{N_j+1}} \prod_{\substack{i=1 \\ i \neq j}}^{N} \left(1 - \frac{(1+S)^{N_i}}{(1+S+T_i)^{N_i}}\right)$$

$$+ \frac{\Gamma}{(1+T_j)^{N_j+1}} \prod_{\substack{i=1 \\ i \neq j}}^{N} \left(1 - \frac{1}{(1+T_i)^{N_i}}\right) = 0, \quad j = 1, 2, ..., N.$$

$$(5.3.15)$$

The threshold multipliers can be determined by solving (5.3.15) along with the constraint

$$1 - \prod_{i=1}^{N} \left(1 - \frac{1}{(1+T_i)^{N_i}}\right) = \alpha. \quad (5.3.16)$$

For the two sensor case, unlike the AND fusion rule, no explicit solutions for T_1 and T_2 are found. Instead, they are determined numerically.

OS-CFAR Processors

In the distributed detection system with OS-CFAR local detectors operating in a homogeneous background, the global detection probability for a given data fusion rule, is maximized by optimizing the parameters of the local detectors simultaneously. For local detector i, these parameters are the scaling factor T_i and the order number k_i. Once again, the Lagrange multiplier technique is employed for system optimization.

5. Distributed Detection with False Alarm Rate Constraints

For a given SNR common to all local detectors and a given set of window sizes at the local detectors, the objective function is formed as

$$J[(T_1, k_1), (T_2, k_2), ..., (T_N, k_N)] = P_D[(T_1, k_1), (T_2, k_2), ..., (T_N, k_N)]$$
$$+ \Gamma\{P_F[(T_1, k_1), (T_2, k_2), ..., (T_N, k_N)] - \alpha\},$$

(5.3.17)

where α is the desired constant false alarm probability at the data fusion center, Γ is the Lagrange multiplier, (T_i, k_i), $i = 1, 2, ..., N$, are the scaling factor and the order number of the sample selected from the ordered reference window at local detector i. The values of P_{Fi} and P_{Di} of the local OS-CFAR detector i are obtained from the expressions in (2.6.22) and (2.6.23), respectively, by replacing N by N_i, T by T_i, and k by k_i.

In the optimization procedure, first, we select a specific set of values for k_i, $i=1, ..., N$, and set the derivatives of the objective function, $J[(T_1, k_1), (T_2, k_2), ..., (T_N, k_N)]$, with respect to T_j, $j = 1, 2, ..., N$, equal to zero to obtain

$$\frac{\partial J[(T_1, k_1), (T_2, k_2), ..., (T_N, k_N)]}{\partial T_j} = 0, \quad j = 1, 2, ..., N.$$

(5.3.18)

This set of equations and the constraint

$$P_F[(T_1, k_1),(T_2, k_2), ..., (T_N, k_N)] = \alpha \quad (5.3.19)$$

yield $(N + 1)$ nonlinear equations with $(N + 1)$ unknowns, where the unknowns are the scaling factors of the local detectors, T_i, $i = 1, 2, ..., N$, and the Lagrange multiplier, Γ. For a given set of values for k_i, $i = 1, 2, ..., N$, the optimum set of T_i, $i = 1, 2, ..., N$, is found by solving equations (5.3.18) and (5.3.19) simultaneously, and the corresponding global probability of detection is computed. For the optimization of the system over k_i, this procedure is repeated for every possible set of values for k_i, $i = 1, ... N$, exhaustively. The set of values for k_i and the corresponding values of optimum T_i, $i = 1, 2, ..., N$, which give the highest global probability of detection, P_D, form the optimum solution for the local detectors.

Next, we present expressions for the objective function for the AND and the OR fusion rules. Detailed derivations of the set of equations to be solved to determine the threshold multipliers are not presented here and can be found in [Une93]. For the AND fusion rule, the objective function is given by

$$J[(T_1, k_1), (T_2, k_2) ..., (T_N, k_N)]$$

$$= \prod_{i=1}^{N} \left[\prod_{j=0}^{k_i-1} (N_i-j)/[N_i-j+T_i/(1+S)] \right]$$

$$+ \Gamma \left\{ \prod_{i=1}^{N} \left[\prod_{j=0}^{k_i-1} (N_i-j)/(N_i-j+T_i) \right] - \alpha \right\}. \quad (5.3.20)$$

For the OR fusion rule,

$$J[(T_1, k_1), (T_2, k_2), ..., (T_N, k_N)]$$

$$= 1 - \prod_{i=1}^{N} \left[1 - \prod_{j=0}^{k_i-1} (N_i-j)/[N_i-j+T_i/(1+S)] \right]$$

$$+ \Gamma \left\{ 1 - \prod_{i=1}^{N} \left[1 - \prod_{j=0}^{k_i-1} (N_i-j)/(N_i-j+T_i) \right] - \alpha \right\}. \quad (5.3.21)$$

Setting the derivatives of the objective function equal to zero, and solving the resulting equations for threshold multipliers under the appropriate false alarm rate constraint, we may obtain the parameters of the distributed OS-CFAR detection system.

Example 5.2

Consider the distributed CFAR detection system consisting of two local detectors and operating with either the OR fusion rule or the AND fusion rule. The system is assumed to operate in a homogeneous background. Reference window sizes at the local detectors are assumed to be $N_1 = 24$ and $N_2 = 16$. The design value of P_F is assumed to be 10^{-4}

202   5. Distributed Detection with False Alarm Rate Constraints

and P_D is maximized for the OR and the AND fusion rules. In Figure 5.7, system performance in terms of P_D versus SNR is presented. As expected, the CA-CFAR system performs better than the OS-CFAR system in a homogeneous background. Optimum order numbers are used in the OS-CFAR detection system. The OR fusion rule performs better than the AND fusion rule.

Performance in Nonhomogeneous Background

In a practical distributed detection system, local detectors are located physically apart from each other. Each local CFAR detector obtains its reference window samples from the neighboring cells of the test cell along the beam direction when observed from the local detector. Each local detector may observe a different nonhomogeneity in its reference window. For example, in the multiple target case, each local detector may observe a different number of interfering targets in its reference window. See Figure 5.8 for an example of this situation with two local detectors. In the clutter-edge case, the reference window of each local detector may be filled with different number of samples from a noise-plus-clutter environment while the remaining parts of the reference windows are filled with samples from a noise-only environment. This situation is illustrated in Figure 5.9 for a system with two local detectors. The performance in both of these nonhomogeneous situations is examined in the next two examples. A more extensive discussion of these issues is available in [Une93, UnV95].

Example 5.3

Consider the two sensor distributed detection system of Example 5.2 with $N_1 = 24$ and $N_2 = 16$. The design value of P_F is assumed to be 10^{-4}. Two multiple target situations are considered. The test cell target and interfering targets are assumed to have the same average power, i.e., $S/I = 1$. In the first scenario, local detector 1 observes four interfering targets whereas local detector 2 does not observe any such targets. In the second scenario, local detector 1 still observes four interfering targets whereas local detector 2 observes two interfering targets. From the performance results presented in Figures 5.10 and 5.11, we observe that the distributed OS-CFAR systems perform much better than the distributed CA-CFAR systems. If one of the local detectors does not

5.3 Distributed CFAR Detection   203

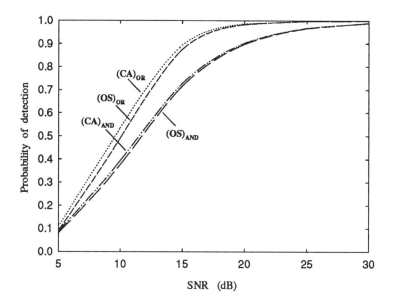

Figure 5.7. Performance of distributed CFAR detection systems in a homogeneous background, $N_1 = 24$, $N_2 = 16$.

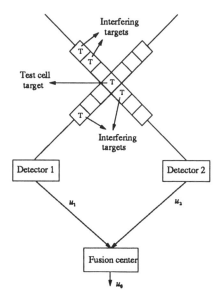

Figure 5.8. A typical distributed CFAR detection system with interfering targets.

204   5. Distributed Detection with False Alarm Rate Constraints

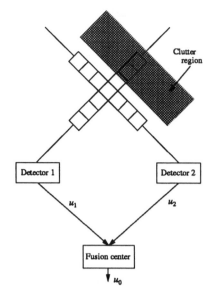

Figure 5.9. A typical distributed CFAR detection system with a clutter edge.

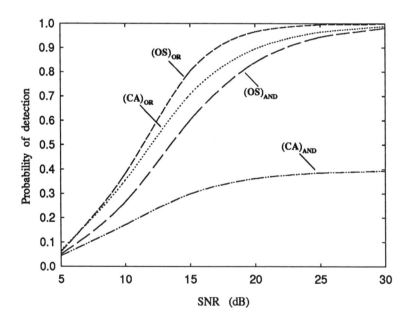

Figure 5.10. Performance of distributed CFAR detection system with interfering targets, first scenario.

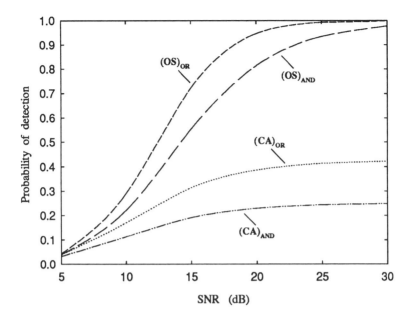

Figure 5.11. Performance of distributed CFAR detection system with interfering targets, second scenario.

have an interfering target in its reference window, then, the distributed CA-CFAR system with the OR fusion rule performs almost as well as the distributed OS-CFAR systems. But, if both of the local detectors have interfering targets in their reference windows, then, the P_D performance of the distributed CA-CFAR system degrades severely.

Example 5.4

Consider the two-sensor distributed detection system of Example 5.2 with $N_1 = N_2 = 24$ operating in a clutter-edge environment. In this problem, maintenance of P_F under the design value is of major concern. Depending upon the location of the test cell with respect to the clutter region, P_F may exceed the design value. Consider the clutter-edge scenario shown in Figure 5.9 and determine system performance in terms of P_F. In this case, the first local detector observes a clutter edge in its reference window whereas the second local detector observes a homogeneous background. It is assumed that clutter enters the reference window from one end, and, until it fills half of the reference window, the test cell remains in the noise-only region. After that, the test cell is also filled with clutter-plus-noise. In Figures 5.12 and 5.13, we show the P_F performance of the system with a clutter-to-noise power ratio of 10 dB. Figure 5.12 shows the case where the test cell is in the noise-only region. In this case, P_F stays below the design value of 10^{-4}. Both CA-CFAR and OS-CFAR systems with the OR fusion rule provide very good P_F control, i.e., the value of P_F is maintained very close to the design value, and this, in turn, keeps the value of P_D high. The performance for the case where the test cell is filled with clutter-plus-noise samples is shown in Figure 5.13. The design value of P_F is exceeded, but OS-CFAR systems provide better P_F control compared with their CA-CFAR counterparts.

5.4 Distributed Detection of Weak Signals

In conventional centralized detection, locally optimum detection procedures are employed to detect weak signals. The slope of the detection probability at zero signal strength (assuming that this slope is not zero) is maximized under a constraint on the probability of false alarm. In this section, we consider the parallel fusion network topology, shown in Figure 3.6, for detecting weak signals. It consists of N local detectors and a fusion center. Based on its own observation, each local detector determines a hard decision u_i, $i = 1, ..., N$, and transmits it to the fusion center. The fusion center combines the local decisions to yield the global decision u_0. The objective for the design of a locally optimum distributed detection system is to maximize the slope of P_D at the fusion center at zero signal strength while constraining P_F to a prespecified value α. From the generalized Neyman–Pearson lemma [Leh86], it follows that, for a one-sided alternative, the locally optimum fusion rule is given by the first derivative of its likelihood ratio evaluated at zero signal strength. For a two-sided alternative in which the first derivative of P_D at zero signal strength is zero for all tests with $P_F = \alpha$, the locally optimum fusion rule is given by the second derivative of the likelihood ratio evaluated at zero signal strength.

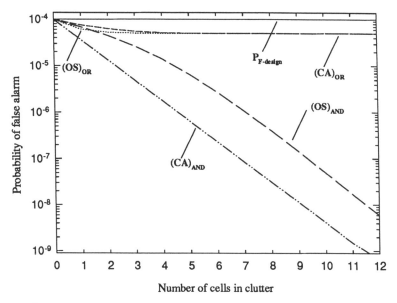

Figure 5.12. Performance of distributed CFAR detection system with clutter edge, test cell in the noise-only region.

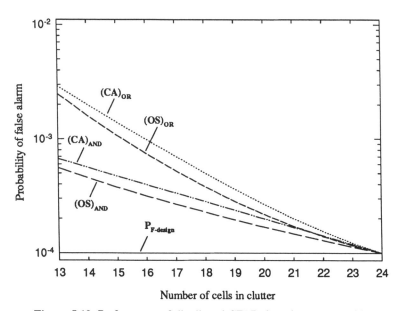

Figure 5.13. Performance of distributed CFAR detection system with clutter edge, test cell in the clutter-plus-noise region.

As before, let y_i, $i = 1, ..., N$, represent the observations at the local detectors. The joint pdf of the observations is denoted by $p(y_1, ..., y_N \mid \theta)$ where θ is a parameter that defines the two hypotheses. The null or noise-only hypothesis H_0 is characterized by $\theta = 0$ whereas the alternative hypothesis H_1 corresponds to $\theta > 0$. The goal is to design the local decision rules and the fusion rule so as to maximize the slope of the detection probability $\partial P_D(\theta)/\partial \theta |_{\theta=0}$ for a given value of P_F, say α. System optimization is carried out using the Lagrange multiplier method. The objective function is expressed as

$$F = \frac{\partial}{\partial \theta} P_D(\theta) |_{\theta=0} - \Gamma (P_F - \alpha)$$

$$= \frac{\partial}{\partial \theta} \left(\sum_u P_{1u} M_u \right) |_{\theta=0} - \Gamma \left(\sum_u P_{1u} F_u - \alpha \right), \quad (5.4.1)$$

where the notation defined in Section 5.2 has been used. After some manipulation, (5.4.1) can be written as

$$F = \sum_u P_{1u} \left(\frac{\partial}{\partial \theta} M_u |_{\theta=0} - \Gamma F_u \right) + \Gamma \alpha. \quad (5.4.2)$$

Clearly, the locally optimum fusion rule that maximizes F is given by

$$P_{1u} = P(u_0 = 1 | u) = \begin{cases} 1, & \text{if } \frac{\partial}{\partial \theta} M_u |_{\theta=0} > \Gamma F_u, \\ \gamma, & \text{if } \frac{\partial}{\partial \theta} M_u |_{\theta=0} = \Gamma F_u, \\ 0, & \text{if } \frac{\partial}{\partial \theta} M_u |_{\theta=0} < \Gamma F_u. \end{cases} \quad (5.4.3)$$

This fusion rule, as stated above, admits the possibility of randomization. As pointed out in Remark 5.2.2, randomized fusion rules need not be considered when the test statistics at the local detectors do not contain point masses of probability.

Next, we determine the local decision rules. We expand F by explicitly summing over u_k:

$$F = \Gamma\alpha + \sum_u P(u_0=1|u)\left[\frac{\partial}{\partial\theta}P(u|H_1)|_{\theta=0} - \Gamma P(u|H_0)\right]$$

$$= \Gamma\alpha + \sum_{u^k}\left\{\frac{\partial}{\partial\theta}[P(u_0=1|u_k=0, u^k)P(u_k=0, u^k|H_1)\right.$$

$$+ P(u_0=1|u_k=1, u^k)P(u_k=1, u^k|H_1)]|_{\theta=0}$$

$$- \Gamma[P(u_0=1|u_k=0, u^k)P(u_k=0, u^k|H_0)$$

$$\left. + P(u_0=1|u_k=1, u^k)P(u_k=1, u^k|H_0)]\right\}, \quad (5.4.4)$$

where u and u^k are as defined earlier. To determine the local decision rule at the kth detector, we express

$$P(u_k, u^k|H_i) = \int_{y_k} P(u_k|y_k)P(u^k|y_k, H_i)p(y_k|H_i)dy_k, i=0, 1. \quad (5.4.5)$$

Substituting from (5.4.5), using the fact that $P(u_k = 0|y_k) = 1 - P(u_k = 1|y_k)$, and rearranging terms, (5.4.4) can be expressed as

$$F = \int_{y_k} P(u_k=1|y_k)\sum_{u^k}\left\{\frac{\partial}{\partial\theta}\left[P(u^k|y_k, H_1)p(y_k|H_1)\right]|_{\theta=0}\right.$$

$$\times\left[P(u_0=1|u_k=1, u^k) - P(u_0=1|u_k=0, u^k)\right]$$

$$- \Gamma[P(u^k|y_k, H_0)p(y_k|H_0)]$$

$$\left.\times\left[P(u_0=1|u_k=1, u^k) - P(u_0=1|u_k=0, u^k)\right]\right\}dy_k - \Gamma',$$

$$(5.4.6)$$

where Γ' includes those terms that are constants as far as the optimization of the kth local detector is concerned. To maximize F, we set

$$P(u_k=1|y_k) = \begin{cases} 1, & \text{if } T_k(y_k) > \Gamma, \\ 0, & \text{if } T_k(y_k) < \Gamma, \end{cases} \quad (5.4.7)$$

where

$$T_k(y_k) = \frac{\sum_{u^k} B^k \frac{\partial}{\partial \theta}\{P(u^k|y_k, H_1)p(y_k|H_1)\}|_{\theta=0}}{\sum_{u^k} B^k P(u^k|y_k, H_0)p(y_k|H_0)}, \quad (5.4.8)$$

and

$$B^k = \left[P(u_0=1|u_k=1, u^k) - P(u_0=1|u_k=0, u^k)\right]. \quad (5.4.9)$$

Here, we have assumed that the event $T_k(y_k)=\Gamma$ occurs with zero probability. Also, we have assumed that the denominator is positive. This assumption is reasonable for all monotonic fusion rules. The test statistic of (5.4.8) simplifies to

$$T_k(y_k) = \frac{\frac{\partial}{\partial \theta} p(y_k|H_1)|_{\theta=0}}{p(y_k|H_0)} + \frac{\sum_{u^k} B^k \frac{\partial}{\partial \theta}\left[P(u^k|y_k, H_1)\right]|_{\theta=0}}{\sum_{u^k} B^k P(u^k|y_k, H_0)}. \quad (5.4.10)$$

It is interesting to observe that the test statistic decomposes into two terms. The first term is the usual locally optimum test statistic when a single sensor is employed. The second term depends on the fusion rule and the performance of the other detectors. A simultaneous solution of the above equations yields the locally optimum distributed detection system. Once again, note that, while determining the decision rule at any detector, decision rules at the other detectors were assumed to be fixed and the optimization methodology employed yields a PBPO solution.

In many practical situations, the first derivative of the marginal pdfs of the observations are zero at $\theta = 0$. In this case, one maximizes the second derivative of the probability of detection at zero signal strength. A similar procedure can be followed to determine the locally optimum fusion rule and local decision rules. The resulting fusion rule is

$$P(u_0 = 1 | u) = \begin{cases} 1, & \text{if } \frac{\partial^2}{\partial \theta^2} M_u |_{\theta = 0} > \Gamma F_u, \\ \gamma, & \text{if } \frac{\partial}{\partial \theta^2} M_u |_{\theta = 0} = \Gamma F_u, \\ 0, & \text{if } \frac{\partial}{\partial \theta^2} M_u |_{\theta = 0} < \Gamma F_u. \end{cases} \quad (5.4.11)$$

The local decision rule is still given by (5.4.7) but with test statistics $T_k(y_k)$ now given as

$$T_k(y_k) = \frac{\frac{\partial^2}{\partial \theta^2} p(y_k | H_1) |_{\theta = 0}}{p(y_k | H_0)} + \frac{\sum_{u^k} B^k \frac{\partial^2}{\partial \theta^2} \left[P(u^k | y_k, H_1) \right] |_{\theta = 0}}{\sum_{u^k} B^k P(u^k | y_k, H_0)}$$

(5.4.12)

where B^k is still as given in (5.4.9).

As pointed out in [BlK92], the decision rules obtained above allow for the observations at the local detectors to be dependent. The second term of the local detector test statistics depends on $P(u^k | y_k, H_i)$, $i=0, 1$, and, therefore, the functional form of the local detector test statistics at different sensors may be different. If, however, the observations at the local detectors are independent, the second term depends on $P(u^k | H_i)$, $i=0, 1$, and the functional form of the local detector test statistics will be identical as shown in [Sri90a]. Of course, the local thresholds may be different.

Example 5.5

Consider a parallel fusion network consisting of two local detectors and a fusion center. We obtain the test statistics given in (5.4.10) and

(5.4.12) for this detection network. As in [BlK92], a change in notation is used to show the explicit dependence of pdfs on θ. The marginal pdf of the observations at local detector k will be denoted as $p(y_k|\theta)$ whereas $p(y_k|y_j, \theta)$ will denote the conditional density of the observation at local detector k given the observation at local detector j. Hypothesis H_0 corresponds to $\theta = 0$ and a positive value of θ corresponds to H_1. Based on this notation, (5.4.10) yields

$$T_1(y_1) = \frac{\frac{\partial}{\partial \theta} p(y_1|\theta)|_{\theta=0}}{p(y_1|\theta=0)} + \frac{\sum_{u_2} B^1 \frac{\partial}{\partial \theta}[P(u_2|y_1, \theta)]|_{\theta=0}}{\sum_{u_2} B^1 P(u_2|y_1, \theta=0)}$$

$$= \frac{\frac{\partial}{\partial \theta} p(y_1|\theta)|_{\theta=0}}{p(y_1|\theta=0)} + \frac{c_1 \int_{T_2 > \Gamma} \frac{\partial}{\partial \theta} p(y_2|y_1, \theta)|_{\theta=0} dy_2}{c_2 + c_1 \int_{T_2 > \Gamma} p(y_2|y_1, \theta=0) dy_2},$$

(5.4.13)

where

$$c_1 = P_{111} - P_{101} - [P_{110} - P_{100}],$$
$$c_2 = P_{110} - P_{100},$$

and

$$P_{ijk} = P(u_0 = i | u_1 = j, u_2 = k).$$

This decision statistic $T_1(y_1)$ is compared with the threshold Γ to make the local decision at the first local detector. Note that $T_1(y_1)$ is coupled with $T_2(y_2)$ by means of the region of integration $T_2(y_2) > \Gamma$. The decision statistic $T_2(y_2)$, obtained similarly is given by

$$T_2(y_2) = \frac{\frac{\partial}{\partial \theta} p(y_2|\theta)|_{\theta=0}}{p(y_2|\theta=0)} + \frac{c_1 \int_{T_1 > \Gamma} \frac{\partial}{\partial \theta} p(y_1|y_2, \theta)|_{\theta=0} dy_1}{c_2 + c_1 \int_{T_1 > \Gamma} p(y_1|y_2, \theta=0) dy_1}.$$

(5.4.14)

Also, the local decision test statistics given by (5.4.12), for the two-sensor case are

$$T_1(y_1) = \frac{\frac{\partial^2}{\partial \theta^2} p(y_1|\theta)|_{\theta=0}}{p(y_1|\theta=0)} + \frac{c_1 \int_{T_2 > \Gamma} \frac{\partial^2}{\partial \theta^2} p(y_2|y_1, \theta)|_{\theta=0} dy_2}{c_2 + c_1 \int_{T_2 > \Gamma} p(y_2|y_1, \theta=0) dy_2},$$

(5.4.15)

and

$$T_2(y_2) = \frac{\frac{\partial^2}{\partial \theta^2} p(y_2|\theta)|_{\theta=0}}{p(y_2|\theta=0)} + \frac{c_1 \int_{T_1 > \Gamma} \frac{\partial^2}{\partial \theta^2} p(y_1|y_2, \theta)|_{\theta=0} dy_1}{c_2 + c_1 \int_{T_1 > \Gamma} p(y_1|y_2, \theta=0) dy_1}.$$

(5.4.16)

When the observations at the local detectors are independent, the second term of the test statistics (5.4.13) to (5.4.16) becomes a constant and can be incorporated in the threshold. In this case, the test statistics reduce to the conventional locally optimum detector statistics.

Next, we consider a specific observation model and treat the detection of a random signal in additive noise. We assume that a single observation is taken at each of the sensors and that the signals observed at the two detectors are the same. The observations at the two sensors are

$$y_1 = \theta s + w_1,$$

and

$$y_2 = \theta s + w_2,$$

where s is a zero mean random signal with unit variance, and w_1 and w_2 are noise samples at the two detectors with pdfs $p_{w_1}(\cdot)$ and $p_{w_2}(\cdot)$, respectively. In this case, the optimum local detector test statistics for the two sensors are given by

$$T_1(y_1) = \frac{p_{w_1}''(y_1)}{p_{w_1}(y_1)} + \frac{p_{w_1}'(y_1)}{p_{w_1}(y_1)} a_1 + b_1, \qquad (5.4.17)$$

and

$$T_2(y_2) = \frac{p_{w_2}''(y_2)}{p_{w_2}(y_2)} + \frac{p_{w_2}'(y_2)}{p_{w_2}(y_2)} a_2 + b_2, \qquad (5.4.18)$$

where

$$a_1 = 2 \frac{c_1}{P_{F2} c_1 + c_2} \left[\int_{T_2 > \Gamma} p_{w_2}'(y_2) dy_2 \right],$$

$$b_1 = \frac{c_1}{P_{F2} c_1 + c_2} \left[\int_{T_2 > \Gamma} p_{w_2}''(y_2) dy_2 \right],$$

$$a_2 = 2 \frac{c_1}{P_{F1} c_1 + c_2} \left[\int_{T_1 > \Gamma} p_{w_1}'(y_1) dy_1 \right],$$

and

$$b_2 = \frac{c_1}{P_{F1} c_1 + c_2} \left[\int_{T_1 > \Gamma} p_{w_1}''(y_1) dy_1 \right].$$

In the test statistics, the first term is the classical locally optimum test statistic of the form p''/p. The second term of the form p'/p attempts to construct an approximate correlation term to account for the presence of the second sensor. The third term is a constant that can be incorporated in the threshold. For a given fusion rule, say the AND or the OR rule, system design involves determining the constants a_1, b_1, a_2, b_2, and Γ. The performance measure for the system is $d^2 P_D(\theta)/d\theta^2 |_{\theta=0}$. For a given noise distribution, numerical techniques can be employed to determine the required constants. Blum and Kassam [BlK92] have considered four different noise pdfs namely Gaussian, Cauchy, sech, and a mixture of Gaussians. They have reported extensive numerical results that are not reproduced here. For the Gaussian case, it was found that the OR fusion rule was the best for all cases considered, with P_F ranging from 0.01 to 0.45. The constants a_1 and a_2 were found to be zero so that the local detector test statistics reduced to the same form as the classical centralized one. For other noise distributions, local detector test statistics differed from the classical centralized ones in many instances.

Notes and Suggested Reading

Distributed Neyman–Pearson detection for the parallel fusion network was considered in [HoV86, Sri86a, TVB87a]. An overview containing additional results is available in [Tsi93a]. Neyman–Pearson detection for other network topologies, such as the serial network and the parallel fusion network with feedback, is discussed in [VTT88, Sri90b]. Distributed CFAR detection was considered in [BaV89]. Further results on the problem are given in [BaV91, EBA92, UnV95, BlK95a]. Distributed detection of weak signals is addressed in [BlK92]. Additional results on this topic are available in [BlK95b, BlK95c].

6
Distributed Sequential Detection

6.1 Introduction

This chapter considers distributed sequential detection problems. In sequential detection, observations are assumed to arrive sequentially at the detectors. As observations continue to arrive, detectors include them in their decision making. Unlike fixed-sample-size detection problems where decisions are made after receiving the entire set of observations, sequential detectors may choose to stop at any time and make a final decision or continue to take additional observations. In this chapter, we will consider a Bayesian formulation of two distributed sequential detection problems for parallel fusion network topologies. In Section 6.2, we consider a parallel fusion network without a fusion center. For simplicity, we restrict our attention to a two-detector network. Sequential tests are implemented at individual detectors. The system is optimized based on a coupled cost assignment. These decisions can still be combined using a fixed fusion rule. In Section 6.3, we consider a parallel fusion network consisting of N peripheral detectors and a fusion center. The local detectors send a sequence of summary messages to the fusion center where a sequential test is implemented. The fusion center makes the decision whether to continue taking additional observations or to stop and make a final decision on the hypothesis present. In general, distributed sequential detection problems are quite complex as is evident from the discussion included in this chapter.

6.2 Sequential Test Performed at the Sensors

Consider the distributed detection system consisting of two detectors and without a fusion center as shown in Figure 3.1. Both detectors observe the same phenomenon and make decisions regarding it by implementing a sequential test. These decisions are available at the local detectors or are transferred to a common remote site. No communication between the detectors is assumed, but they are used to achieve a certain common systemwide goal. A coupled cost assignment appropriate for the common goal is employed. Consider a Bayesian formulation of the binary hypothesis testing problem for hypotheses H_0 and H_1 with a priori probabilities P_0 and P_1. Let y_i^t, $i=1, 2$, denote the observation at DM i at time t. Observations at the two detectors are assumed to be conditionally independent. DM i takes observations until its stopping time τ_i and reaches its decision u_i based on its observations until τ_i. Decision policy at DM i is denoted by $\gamma_i(\cdot)$ and involves the determination of the decision u_i as well as the stopping time τ_i. As in Section 3.2, the common decision cost function is defined as C_{ijk} and it represents the cost of DM 1 deciding H_i, DM 2 deciding H_j when H_k is present. These costs C_{ijk} are not decomposable, i.e., $C_{ijk} \neq C_{ik} + C_{jk}$, otherwise the problem decomposes into two independent problems at the two sensors. Also, the following reasonable assumptions regarding costs with respect to u_2 are made

$$C_{0j1} \geq C_{1j1}, \quad C_{1j0} \geq C_{1j1},$$
$$C_{1j0} \geq C_{0j0}, \quad C_{0j1} \geq C_{0j0}.$$

Similar assumptions with respect to u_1 are made. We also associate a positive cost c with each observation taken by a local detector.

Consider the Bayesian optimization problem where the objective is to derive decision policies γ_1 and γ_2 that minimize an average cost. The cost is defined to be a linear combination of the stopping times and the cost of decision making as in [TeH87]. Another cost function that is a linear combination of the maximum of two stopping times and the cost of decision making has been considered in [Vee92a] and is not considered here. The problem that we wish to solve is

$$\min_{\gamma_1 \in \Gamma_1, \gamma_2 \in \Gamma_2} E\{c\tau_1 + c\tau_2 + C_{u_1,u_2,k}\}, \quad (6.2.1)$$

where Γ_1 and Γ_2 denote the sets of all decision policies at DM 1 and DM 2.

We employ the person-by-person optimization (PBPO) methodology for system optimization. First, we assume that $\gamma_2 \in \Gamma_2$ is fixed and derive $\gamma_1 \in \Gamma_1$. The problem at DM 1 is to determine the decision u_1 and the stopping time τ_1 so that the average cost is minimized, i.e.,

$$\min_{u_1 \in \{0,1\}, \tau_1} E\{c\tau_1 + C_{u_1,u_2,k} | Y_{\tau_1,1}\}, \quad (6.2.2)$$

where $Y_{\tau_1,1}$ represents the set of observations at DM 1 up to time τ_1, i.e., $Y_{\tau_1,1} = \{y_1^1, y_1^2, ..., y_1^{\tau_1}\}$. This problem can be solved by backward induction. First, we consider the finite-horizon problem, i.e., the stopping times τ_1 and $\tau_2 \leq T$ for some finite T. Then, we consider the infinite-horizon problem and let $T \to \infty$.

In order to solve this problem, we introduce the statistic

$$\pi_t = P(H_0 | Y_{t,1}). \quad (6.2.3)$$

Define

$$q(y_1^{t+1} | \pi_t) = \pi_t p(y_1^{t+1} | H_0) + (1-\pi_t) p(y_1^{t+1} | H_1) \quad (6.2.4)$$

and

$$\phi(\pi_t, y_1^{t+1}) = \frac{\pi_t p(y_1^{t+1} | H_0)}{q(y_1^{t+1} | \pi_t)}, \quad (6.2.5)$$

where $p(\cdot | H_j)$, $j=0, 1$, denotes the conditional density of the observations. Using Bayes rule,

$$p(y_1^{t+1}|Y_{t,1}) = q(y_1^{t+1}|\pi_t) , \qquad (6.2.6)$$

and

$$\pi_{t+1} = \phi(\pi_t, y_1^{t+1}) . \qquad (6.2.7)$$

The value function for the dynamic programming problem at time t, $J_t^T(\pi_t)$, is given by

$$J_T^T(\pi_T) = \min\{G_0(\gamma_2)\pi_T + K_0(\gamma_2),\ G_1(\gamma_2)\pi_T + K_1(\gamma_2)\} , \qquad (6.2.8)$$

$$J_t^T(\pi_t) = \min\{G_0(\gamma_2)\pi_t + K_0(\gamma_2),\ G_1(\gamma_2)\pi_t + K_1(\gamma_2),\ c + A_t^T(\pi_t)\},$$
$$t = 1, 2, ..., T-1 , \qquad (6.2.9)$$

where

$$K_i(\gamma_2) = \sum_{u_2} P(u_2|H_1)C_{iu_2 1},\ i = 0, 1,$$

$$G_i(\gamma_2) = \sum_{u_2} P(u_2|H_0)C_{iu_2 0} - K_i(\gamma_2),\ i = 0, 1,$$

and

$$A_t^T(\pi_t) = \int J_{t+1}^T\bigl(\phi(\pi_t, y_1^{t+1})\bigr) q(y_1^{t+1}|\pi_t) dy_1^{t+1} .$$

Examine the three terms inside the $\min\{\cdot\}$ function in (6.2.9). The first term is the cost of stopping at time t and deciding H_0, the second term represents the cost of stopping at time t and deciding H_1, and the third term represents the cost of continuing at time t. It is optimal to stop at time t if, and only if, the cost of continuing is greater than or equal to the cost of stopping at time t and deciding H_0 or H_1, i.e.,

$$\min_{i \in \{0,1\}} \{G_i(\gamma_2)\pi_t + K_i(\gamma_2)\} \leq c + A_t^T(\pi_t) . \quad (6.2.10)$$

It has been shown in [TeH87] that $J_t^T(\pi_t)$ is a nonnegative concave function of π_t, $t=1, 2, ..., T$. Also, the following inequalities at $\pi_t=0$ and $\pi_t=1$ were shown to hold for $t=1, 2, ..., T-1$:

$$\min_{i \in \{0,1\}} \{G_i(\gamma_2)\pi_t + K_i(\gamma_2)\}|_{\pi_t=0} < c + A_t^T(\pi_t) , \quad (6.2.11)$$

and

$$\min_{i \in \{0,1\}} \{G_i(\gamma_2)\pi_t + K_i(\gamma_2)\}|_{\pi_t=1} < c + A_t^T(\pi_t) . \quad (6.2.12)$$

In addition, $J_t^T(\cdot)$ and $A_t^T(\cdot)$ were shown to be monotonically nondecreasing in t. Based on these results, when the decision policy at DM 2 is fixed, the optimal stopping rule at DM 1 can be described by thresholds m_T, $\zeta_1, \xi_1, ..., \zeta_{T-1}, \xi_{T-1}$. The optimal stopping time at DM 1 is given by

$$\tau_1 = \min\{t: \pi_t \leq \zeta_t \text{ or } \pi_t \geq \xi_t\} , \quad (6.2.13)$$

where the threshold m_T is obtained by solving

$$G_0(\gamma_2)m_T + K_0(\gamma_2) = G_1(\gamma_2)m_T + K_1(\gamma_2),$$

the thresholds $\zeta_1, ..., \zeta_{T-1}$ are determined by solving

$$G_1(\gamma_2)\zeta_t + K_1(\gamma_2) = c + A_t^T(\zeta_t),$$

and $\xi_1, ..., \xi_{T-1}$ are determined by solving

$$G_0(\gamma_2)\xi_t + K_0(\gamma_2) = c + A_t^T(\xi_t) .$$

These relationships follow directly from (6.2.8) and (6.2.9). The main idea is that DM 1 stops as soon as the cost of stopping and making a decision becomes less than or equal to the cost of taking additional observations.

Now consider the infinite-horizon problem where the stopping time is not limited by T and can take any value. We define the following limit:

$$J_t(\pi_t) = \lim_{T \to \infty} J_t^T(\pi_t) = \inf_T J_t^T(\pi_t) = J(\pi) . \quad (6.2.14)$$

The value function for the infinite-horizon problem is defined as

$$J(\pi) = \min\{G_0(\gamma_2)\pi + K_0(\gamma_2),\ G_1(\gamma_2)\pi + K_1(\gamma_2),\ c + A(\pi)\} , \quad (6.2.15)$$

where

$$A(\pi) = \lim_{T \to \infty} A_t^T(\pi_t) .$$

The optimal stopping time at DM 1 is given by

$$\tau_1 = \min\{t : \pi \le \zeta \text{ or } \pi \ge \xi\} , \quad (6.2.16)$$

where ζ and ξ are determined by solving

$$G_1(\gamma_2)\zeta + K_1(\gamma_2) = c + A(\zeta) .$$

and

$$G_0(\gamma_2)\xi + K_0(\gamma_2) = c + A(\xi) .$$

The optimal stopping rule at DM 1 can be described in terms of two thresholds ζ and ξ. Similarly, for a fixed decision policy at DM 1, the decision policy at DM 2 can be specified in terms of two thresholds. Thus, we have four coupled equations with four unknown, time-invariant thresholds. A simultaneous solution of these equations yields the PBPO solution.

For the finite-horizon problem with horizon T, the PBPO thresholds are time varying. There are four thresholds for each t, $t=1, 2, ..., T-1$

and two thresholds at $t=T$. Thus, a total of $(4T-2)$ coupled equations need to be solved to determine the $(4T-2)$ PBPO thresholds.

Example 6.1

In this example, we obtain some specific results for the infinite-horizon problem. Assume that the observations at the two sensors are identically distributed and are Gaussian with unit variance. Under H_0, the mean is zero whereas it is one under H_1. The cost assignment is assumed to be as given in (3.2.20). Here, we employ the approximations developed in Section 2.5 to determine the decision policies at the two detectors and to evaluate their performance. Let S_{1j} and S_{0j}, $j=1, 2$, denote the average sample sizes at DM j under H_1 and H_0 as given in (2.5.29) and (2.5.30). For the Gaussian observation model with the above parameters, we may write

$$S_{1j} = 2\left[(1-P_{Mj})\log\frac{1-P_{Mj}}{P_{Fj}} + P_{Mj}\log\frac{P_{Mj}}{1-P_{Fj}}\right], \quad j=1, 2,$$

$$S_{0j} = -2\left[P_{Fj}\log\frac{1-P_{Mj}}{P_{Fj}} + (1-P_{Fj})\log\frac{P_{Mj}}{1-P_{Fj}}\right], \quad j=1, 2,$$

where P_{Fj} and P_{Mj} represent the probabilities of false alarm and miss at DM j. Recall that two thresholds are employed at each detector and these probabilities are defined accordingly.

It is desired to minimize the average cost given by

$$E\{c\tau_1 + c\tau_2 + C_{u_1 u_2 k}\} = c[P_1(S_{11}+S_{12}) + P_0(S_{01}+S_{02})]$$
$$+ [P_0(1-P_{F1})P_{F2} + P_0 P_{F1}(1-P_{F2}) + P_1 P_{M1}(1-P_{M2})$$
$$+ P_1(1-P_{M1})P_{M2} + kP_0 P_{F1} P_{F2} + kP_1 P_{M1} P_{M2}]. \quad (6.2.17)$$

The average cost is a function of four thresholds and can be minimized to determine the PBPO values of the thresholds. Here, we constrain the pair of thresholds at the two detectors to be identical, so that only two thresholds need to be determined. It should be pointed out that the thresholds obtained by the minimization of (6.2.17) are employed when the log likelihood ratio is used as a sufficient statistic at the local detectors instead of π, as discussed earlier in this section. Because the log likelihood ratio is equal to $\log((1-\pi)/\pi)$, one may easily obtain the required thresholds when the test is implemented in terms of π.

Next, we present some numerical results in Figures 6.1 to 6.5. The probabilities of false alarm and miss at the two local detectors as a function of the prior probability P_0 are shown in Figures 6.1 and 6.2. Three values of c are employed, namely, $c=0.1$, 0.2, and 0.3; k is set equal to 4. Corresponding thresholds are shown in Figure 6.3. As expected, the probability of false alarm decreases and the probability of miss increases as P_0 increases. The probability of false alarm and the corresponding thresholds as a function of k for $c=0.1$, 0.2, and 0.3 are shown in Figures 6.4 and 6.5. The probabiliy of miss is equal to the probability of false alarm in this example. P_0 is set equal to 0.5. In this case, both types of probability of error decrease as k increases. This is because detectors become more conservative and make fewer errors when double errors are penalized more heavily.

In some applications, sequential tests are implemented at the local detectors, and their decisions are employed at a common site to attain some systemwide objective, say, to trigger a certain action. In this case a decision is to be made whether to trigger the action or not. As such, this requires fusion of incoming local decisions. Unlike fixed-sample-size detection problems, local decisions may arrive at the fusion center

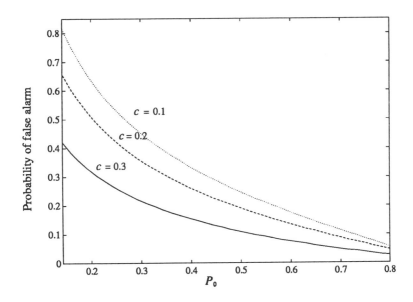

Figure 6.1. Probability of false alarm as a function of P_0.

224   6. Distributed Sequential Detection

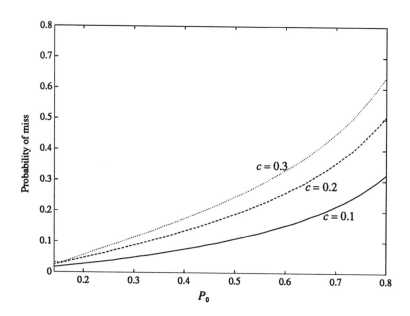

Figure 6.2. Probability of miss as a function of P_0.

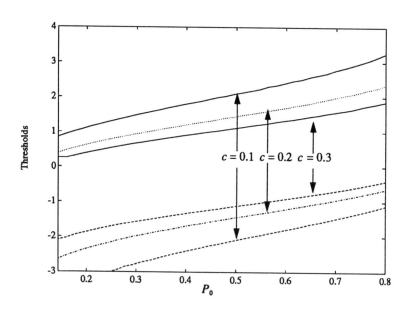

Figure 6.3. Thresholds as a function of P_0.

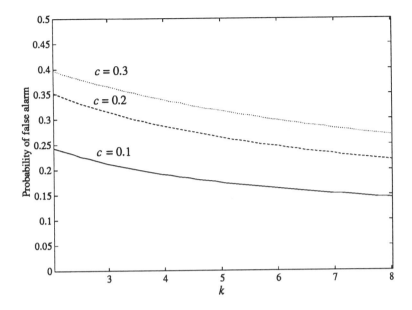

Figure 6.4. Probability of false alarm as a function of k.

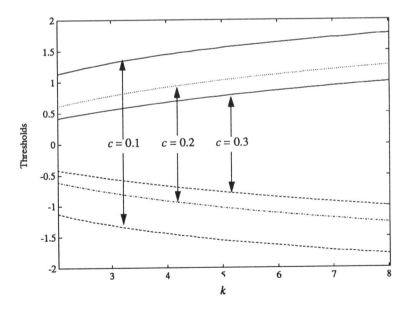

Figure 6.5. Thresholds as a function of k.

at different times depending upon the stopping times of individual detectors. Fusion of decisions in this context has been investigated in [All89]. One option, of course, is to wait until all local decisions have arrived, and, then, employ fusion rules, such as the AND or the OR rule. This approach, however, is counter to the philosophy of sequential testing. One particularly simple scheme is to accept the first incoming local decision and declare it to be the final decision. In this case, the stopping time for the entire system is the minimum of the stopping times of the individual local detectors. This and other similar schemes are analyzed in [AlV88, AlV89] and are not presented here.

6.3 Sequential Test Performed at the Fusion Center

In this section, we consider the parallel fusion network topology of Figure 3.6 consisting of N detectors and a fusion center. All local detectors observe a common phenomenon and send a sequence of local decisions to the fusion center. A sequential test is implemented at the fusion center that makes the global decision for the binary hypothesis testing problem under consideration. Let y_i^t again denote the observation at DM i at time t. These observations are assumed to be independent in time and identically distributed. Also, observations at different sensors are assumed to be conditionally independent. At time t, each DM i transmits its local decision u_i^t to the fusion center. The fusion center is faced with a conventional sequential hypothesis testing problem wherein it has to decide on a stopping time τ and to make the global decision u_0^τ. Bayesian formulation of the problem is considered. Decision policy at the fusion center $\gamma_0(\cdot)$ and those at the local detectors $\gamma_i(\cdot)$, $i=1 \ldots, N$, are determined so that they minimize the average system cost. System cost includes the cost of taking additional observations and the cost of making the global decision. Let c be a positive cost associated with each time step that observations are taken. Also, let C_{ij} be the cost of the global decision being H_i when H_j is present. The problem that we wish to solve is

$$\min_{\{\gamma_0, \gamma_1, \ldots, \gamma_N\}} E\left\{c\tau + C_{ij}\right\}, \qquad (6.3.1)$$

i.e., we wish to determine decision policies at the fusion center and at the local detectors over the set of all decision policies (infinite-horizon) that minimize the average cost. In the rest of this section, we will assume $C_{00}=C_{11}=0$.

The structure of the local decision rules depends upon the information available at the local detectors and the information that the local decisions are allowed to depend upon. We assume the existence of feedback channels from the fusion center to the local detectors, so that the past local decisions from all local detectors can be assumed to be available at each local detector. This is in contrast to the parallel fusion network with feedback considered in Chapter 4, where only past global decisions were assumed to be available at the local detectors. Several cases can be considered where the local decision of DM i at time t, u_i^t, depends upon different types of information.

Case A: u_i^t depends on the current local observation at DM i, y_i^t.

Case B: u_i^t depends on the current local observation at DM i, y_i^t, as well as its past observations $y_i^1, ..., y_i^{t-1}$.

Case C: u_i^t depends on y_i^t and past decisions $u_i^1, ..., u_i^{t-1}$ of DM i.

Case D: u_i^t depends on the current observation y_i^t, the past observations $y_i^1, ..., y_i^{t-1}$, and past decisions from all local detectors $u_1^1, ..., u_1^{t-1}, u_2^1, ..., u_2^{t-1}, ..., u_N^1, ..., u_N^{t-1}$.

Case E: u_i^t depends on the current observation y_i^t and past decisions from all local detectors $u_1^1, ..., u_1^{t-1}, ..., u_N^1, ..., u_N^{t-1}$.

It has been shown in [VBP93] that PBPO local decision rules are likelihood ratio tests in Cases A, C and E but not in cases B and D. Here, we briefly discuss Case E. Detailed derivations and proofs are available in [Vee92a]. Specifically, we assume

$$u_i^t = \gamma_i^t(y_i^t, I^{t-1}) , \qquad (6.3.2)$$

where $\gamma_i^t(.,.)$ is the local decision rule at DM i at time t, and

$$I^{t-1} = \left\{ u_1^1, ..., u_1^{t-1}, u_2^1, ..., u_2^{t-1}, ..., u_N^1, ..., u_N^{t-1} \right\} .$$

Let $\gamma^t = \{\gamma_1^t, ..., \gamma_N^t\}$ denote the set of all local decision rules at time t.

First we discuss the finite-horizon problem in which the stopping time τ is assumed to be less than or equal to some finite T. Let X_T denote the set of all local observations up to time T, i.e.,

$$X_T = \{Y_{T,1}, Y_{T,2}, ..., Y_{T,N}\},$$

where $Y_{T,i}$ is as defined before. The cost of the sequential test is a function of the set of all local decisions up to time T, I^T, the decision policy γ_0 of the fusion center, and the hypothesis present. Let $G(I^T, H_j)$ denote this cost when H_j is present. It obviously depends upon the set of local decision rules. The finite-horizon problem can be stated as

$$\min_{\gamma_0, \gamma^1, ..., \gamma^T} E_{X_T, H_j} G(I^T, H_j). \tag{6.3.3}$$

We obtain the local decision rules based on the PBPO methodology. Let us determine the decision rule for DM i at some specific time t, i.e., we determine γ_i^t. We assume that the fusion rule and all the other local decision rules (including decision rules for DM i at other times) are fixed. In this case, the cost to be minimized is a function of γ_i^t. Let $\Re(\gamma_i^t)$ denote this cost which can be expressed as

$$\Re(\gamma_i^t) = E_{I^{t-1}, y_i^t, H_j} \Big\{ E_{y_i[t+1,T], y_1[t,T],...,y_{i-1}[t,T], y_{i+1}[t,T],...,y_N[t,T] | H_j}$$

$$\{G(\gamma_i^t(y_i^t, I^{t-1}), I^{t-1}, u_1^t, ..., u_{i-1}^t, u_{i+1}^t, ...u_N^t,$$

$$u_1[t+1, T], ..., u_N[t+1, T], H_j) \}\Big\}, \tag{6.3.4}$$

where $y_i[a,b] = \{y_i^a, ..., y_i^b\}$ and $u_i[a,b] = \{u_i^a, ..., u_i^b\}$. The above expression is a consequence of the conditional independence assumption and the fact that the local decision rules up to time $t-1$ are fixed. Denoting the inner expectation by $K(.,.,.)$,

$$\Re(\gamma_i^t) = E_{I^{t-1},y_i^t,H_j}\{K(\gamma_i^t(y_i^t, I^{t-1}), I^{t-1}, H_j)\}$$

$$= E_{I^{t-1},y_i^t}\{P(H_0|y_i^t, I^{t-1})K(\gamma_i^t(y_i^t, I^{t-1}), I^{t-1}, H_0)$$

$$+P(H_1|y_i^t, I^{t-1})K(\gamma_i^t(y_i^t, I^{t-1}), I^{t-1}, H_1)\} . \quad (6.3.5)$$

Minimization of $\Re(\cdot)$ with respect to γ_i^t is equivalent to the minimization of the quantity inside the expectation. Therefore, the PBPO decision rule (when it exists) is given by

$$\gamma_i^t(y_i^t, I^{t-1}) = \arg \min_{u_i^t \in \{0,1\}} \{P(H_0|y_i^t, I^{t-1})K(u_i^t, I^{t-1}, H_0)$$

$$+P(H_1|y_i^t, I^{t-1})K(u_i^t, I^{t-1}, H_1)\} . \quad (6.3.6)$$

This decision rule can be shown to be a likelihood ratio test given by

$$\gamma_i^t(y_i^t, I^{t-1}) = \arg \min_{u_i^t \in \{0,1\}} \{(1-\pi_{t-1})K(u_i^t, I^{t-1}, H_1)\Lambda(y_i^t)$$

$$+\pi_{t-1}K(u_i^t, I^{t-1}, H_0)\} , \quad (6.3.7)$$

where

$$\pi_t = P(H_0|I^t) ,$$

$$\Lambda(y_i^t) = \frac{p(y_i^t|H_1)}{p(y_i^t|H_0)} .$$

The above rule assumes a finite value of the likelihood ratio.

This optimization problem can be solved by backward induction. A sufficient statistic for a dynamic programming solution is the posterior probability π_t, $t=0, ..., T$. Recursion for π_t is given by

$$\pi_{t+1} = \frac{\pi_t P(u_1^{t+1}, \ldots, u_N^{t+1} | H_0, I^t)}{P(u_1^{t+1}, \ldots, u_N^{t+1} | I^t)} \qquad (6.3.8)$$

with $\pi_0 = P_0$. Dynamic programming equations are obtained in the following manner. The value function for this problem at time t, $0 \le t \le T$, is a function of I^t and is denoted by $J_t^T(I^t)$. For $t=T$,

$$J_T^T(I^T) = \min\{(1-\pi_T)C_{01}, \pi_T C_{10}\}, \qquad (6.3.9)$$

where the first term inside the minimum is the conditional expected cost of deciding H_0 given I^T and the second term is the corresponding cost of deciding H_1. For $0 \le t < T$, the value function is given by

$$J_t^T(I^T) = \min\left\{(1-\pi_t)C_{01}, \pi_t C_{10}, c + \inf_q E\{J_{t+1}^T(I^{t+1}) | I^t\}\right\},$$

$$(6.3.10)$$

where q is the joint distribution of the observation vector conditioned on each hypothesis. The third term inside the minimum represents the minimum expected cost of continuing conditioned on I^t and is denoted by $A_t^T(\pi_t)$.

It has been shown in [Vee92a] that both $J_t^T(\pi)$ and $A_t^T(\pi)$ are nonnegative concave functions of π for $\pi \in [0,1]$. They are monotonically nondecreasing in t, i.e., for each $\pi \in [0,1]$,

$$J_t^T(\pi) \le J_{t+1}^T(\pi), \qquad 0 \le t \le T-1,$$

and

$$A_t^T(\pi) \le A_{t+1}^T(\pi), \qquad 0 \le t \le T-2.$$

Furthermore,

$$A_t^T(0) = A_t^T(1) = 0.$$

It is assumed that

$$A_{T-1}^T\left(\frac{C_{01}}{C_{10}+C_{01}}\right) \le \frac{C_{01}C_{10}}{C_{01}+C_{10}}. \quad (6.3.11)$$

Then, the optimal finite-horizon policy at the fusion center has the following form

$$\begin{array}{ll} \text{Accept } H_0, & \text{if } \pi_t \ge \xi_{St}^T, \\ \text{Accept } H_1, & \text{if } \pi_t \le \zeta_{St}^T, \\ \text{Continue}, & \text{otherwise}. \end{array} \quad (6.3.12)$$

The thresholds are specified by the relationships

$$C_{01}(1-\zeta_t^T) = c + A_t^T(\zeta_t^T), \quad (6.3.13)$$

and

$$C_{10}\xi_t^T = c + A_t^T(\xi_t^T). \quad (6.3.14)$$

The thresholds $\{\xi_t^T\}$ form a nonincreasing sequence whereas $\{\zeta_t^T\}$ is a nondecreasing sequence. If the inequality (6.3.11) is not satisfied, ξ_t^T and ζ_t^T become identically equal to $C_{01}/(C_{01} + C_{10})$ for all t greater than some $t' < t$. In that case, the finite horizon simply reduces from T to t'.

In the infinite-horizon version of the problem, stopping time need not be less than some finite T and the restriction on τ can be removed. We solve the problem by letting $T \to \infty$. It can be seen that limits of $J_t^T(\cdot)$ and $A_t^T(\cdot)$, as $T \to \infty$, are well defined and are independent of t. Let $J(\cdot)$ and $A_J(\cdot)$ denote these limits. The value function for the infinite-horizon problem is

$$J(\pi) = \min\{(1-\pi)C_{01}, \pi C_{10}, c + A_J(\pi)\}. \quad (6.3.15)$$

In this case also, $J(\pi)$ and $A_J(\pi)$ are nonnegative concave functions of π, $\pi \in [0,1]$ and

$$J(0) = J(1) = A_J(0) = A_J(1) = 0. \quad (6.3.16)$$

Under the assumption that

$$J\left(\frac{C_{01}}{C_{01}+C_{10}}\right) < \frac{C_{01}C_{10}}{C_{01}+C_{10}}, \qquad (6.3.17)$$

the optimum fusion center policy is of the form

$$\begin{array}{ll} \text{Accept } H_0, & \text{if } \pi_t \geq \xi, \\ \text{Accept } H_1, & \text{if } \pi_t \leq \zeta, \\ \text{Continue}, & \text{otherwise}. \end{array} \qquad (6.3.18)$$

The thresholds are given by the relationships

$$C_{01}(1-\zeta) = c + A_J(\zeta), \qquad (6.3.19)$$

and

$$C_{10}\xi = c + A_J(\xi). \qquad (6.3.20)$$

When (6.3.17) is not satisfied, the optimum fusion center policy is to ignore all the data it receives and base its decision solely on a priori probabilities, i.e., on P_0.

Notes and Suggested Reading

Distributed sequential detection where the sequential test is performed at the sensors was studied in [TeH87]. It is further examined in [Vee92a]. Fusion rules for such systems are derived in [AlV88, AlV89]. Systems where the sequential test is implemented at the fusion center are studied in [VBP93]. Some earlier results on this problem are available in [HaR89], [Vee92b]. A system in which sequential tests are implemented at both the sensors and at the fusion center is described in [Hus94].

7
Information Theory and Distributed Hypothesis Testing

7.1 Introduction

Information theory was developed to determine the fundamental limits on the performance of communication systems. Detection theory involves an application of statistical decision theory to the problem of determining the presence or absence of signals in noise. Both of these theories deal with the communication problem. The relationship between information theory and conventional (centralized) detection theory has been discussed in the literature. Kullback [Kul59] discussed the use of discrimination to study hypothesis testing problems. Middleton [Mid60] employed cost functions based on information theory for optimizing signal detection systems. Csiszar et al. [CsL71] and Blahut [Bla74] have formulated the detection problem as a coding problem and have carried out an asymptotic analysis of detection systems based on the error exponent function. In this chapter, we briefly present some extensions of the above work for the distributed detection problem. In Section 7.2, we discuss the design of distributed detection systems based on an information theoretic cost function. In Section 7.3, we present a brief summary of some asymptototic results on the performance of distributed detection systems.

7.2 Distributed Detection Based on Information Theoretic Criterion

Distributed hypothesis testing problems based on a variety of system optimization criteria have been discussed. For example, in the Bayesian formulation, a fixed cost was assigned to each possible course of action and, then, the average cost based on prior probability knowledge was minimized. In applications where costs are available and are meaningful, Bayesian cost formulation provides an excellent choice for system optimization. However, this may not be the case in all applications. In some applications, our interest may be the amount of information that we are able to transfer; for example, in telephone channels we are concerned with the amount of information transmitted rather than the nature of the information itself. In such situations, entropy-based cost functions may be more meaningful for system optimization. Middleton [Mid60] examined the problem of entropy-based detection in great detail. Middleton [Mid60] and Gabrielle [Gab66] designed an optimum decision system where they minimized the equivocation (or information loss) between the input and the output of the detection system. In this section, we examine the correspondence between the information transmission problem and the detection problem. We also consider the design of parallel fusion networks using entropy-based cost functions.

Correspondence Between Detection Theory and Information Theory

Consider the block diagrams of a conventional binary detection system and a binary communication channel shown in Figure 7.1. The source in the detection problem of Figure 7.1(a) can be viewed as the information source in the information transmission problem. The part of the detection system enclosed by broken lines corresponds to the channel of the information transmission system. The decisions in Figure 7.1(a) may be looked at as the channel output shown in Figure 7.1(b). In the detection problem, the input is a random variable H which may assume one of two values 0 or 1 where $H = i$, $i = 0, 1$, corresponds to the hypothesis H_i being present. The output is the decision random variable u which may again assume the value 0 or 1. The probability of detection P_D, the probability of miss P_M, and the probability of false alarm P_F in the detection problem can be interpreted as the transition probabilities for the information transmission problem as indicated in Figure 7.2.

7.2 Distributed Detection Based on Information Theoretic Criterion 235

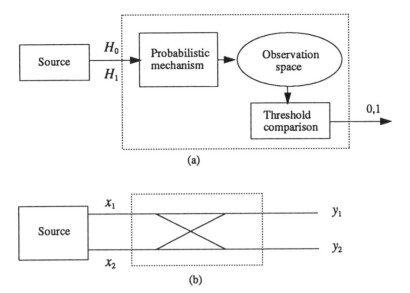

Figure 7.1. Correspondence between signal detection and information transmission systems.

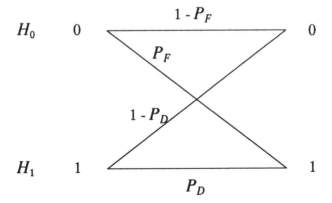

Figure 7.2. Channel model for the signal detection problem.

Let the objective of the detection problem be the minimization of equivocation between the input and the output. In this case, an average cost that is a function of the a posteriori probability of H, given u, is minimized. We minimize the conditional entropy of H, given u, i.e.,

$$h(H|u) = E\{\log[1/P(H|u)]\} . \quad (7.2.1)$$

Given the a priori probabilities P_0 and P_1, our goal is to obtain the decision rule that minimizes $h(H|u)$. This objective is equivalent to the maximization of mutual information $I(H;u)$ because

$$I(H;u) = h(H) - h(H|u) , \quad (7.2.2)$$

and $h(H)$ is constant when P_0 and P_1 are known. It is shown in [HoV89a] that, given the a priori probabilities P_0 and P_1, for each value of P_F (or P_D), the minimum mutual information $I_{min}(H;u)$ is achieved at the point where $P_D = P_F$. Because $I(H;u)$ is a convex upward function of P_D for a given value of P_F and the minimum value of $I(H;u)$ is achieved by choosing $P_D = P_F$, for values of P_D such that $P_D \geq P_F$, $I(H;u)$ is an increasing function of P_D. Therefore, for a fixed value of P_F, maximization of P_D also maximizes $I(H;u)$, i.e., the detector that maximizes $I(H;u)$ for a given P_F is equivalent to the Neyman–Pearson detector for the same value of P_F, and the operating point lies on the ROC of the optimum Bayesian detection system. Hence, the detector that maximizes $I(H;u)$ can be implemented in terms of a likelihood ratio test. Computation of the threshold (information-optimal threshold) requires the knowledge of a priori probabilities but not the set of costs as required in a Bayesian test. It should, however, be pointed out that for every information-optimal threshold, a set of costs exists, so that an equivalent Bayesian test can be determined. But, for different information-optimal thresholds, the relationship among costs for the equivalent Bayesian test changes.

Threshold Computation

Next, we determine the information-optimal threshold for the minimum equivocation detection (MED) problem. A priori probabilities P_0 and P_1 and the conditional densities $p(y|H_j)$, $j=0, 1$, are assumed to be known. The optimum decision rule that maximizes the mutual information $I(H;u)$ or, equivalently, minimizes the equivocation $h(H|u)$ is implemented as a threshold detector with the following likelihood ratio test:

$$\Lambda(y) = \frac{p(y|H_1)}{p(y|H_0)} \underset{H_0}{\overset{H_1}{\gtrless}} t. \qquad (7.2.3)$$

Mutual information $I(H;u)$ is given by

$$I(H;u) = \sum_H \sum_u P(H, u) \log\left\{\frac{P(H|u)}{P(H)}\right\}. \qquad (7.2.4)$$

The a posteriori probabilities are

$$P(u=0) = P_0(1-P_F) + (1-P_0)(1-P_D)$$
$$= \alpha_0, \qquad (7.2.5)$$

and

$$P(u=1) = P_0 P_F + (1-P_0) P_D$$
$$= \alpha_1. \qquad (7.2.6)$$

Substituting appropriate probabilities and simplifying, $I(H;u)$ may be expressed as

$$I(H;u) = P_0(1-P_F)\log(1-P_F) - P_0(1-P_F)\log\alpha_0$$
$$-(1-P_0)(1-P_D)\log\alpha_0 + P_0 P_F \log P_F$$
$$-(1-P_0)P_D\log\alpha_1 + (1-P_0)P_D\log P_D$$
$$+(1-P_0)(1-P_D)\log(1-P_D) - P_0 P_F \log\alpha_1 \ . \qquad (7.2.7)$$

Taking the derivative of $I(H; u)$ with respect to P_F, setting

$$\frac{\partial P_D}{\partial P_F} = t, \qquad (7.2.8)$$

and rearranging the result,

$$\frac{\partial I(H;u)}{\partial P_F} = -P_0 \log\frac{1-P_F}{P_F} + \left[P_0 + t(1-P_0)\right]\log\alpha_0$$
$$-t(1-P_0)\log\frac{1-P_D}{P_D} - \left[P_0 + t(1-P_0)\right]\log\alpha_1 \ . \qquad (7.2.9)$$

Setting this derivative equal to zero and solving for t, we obtain

$$t = \frac{-P_0\left\{\log(\alpha_0/\alpha_1) - \log[(1-P_F)/P_F]\right\}}{(1-P_0)\left\{\log(\alpha_0/\alpha_1) - \log[(1-P_D)/P_D]\right\}} \ . \qquad (7.2.10)$$

The threshold t can be determined by solving (7.2.10). Note that the right hand side is also a function of t because P_D and P_F depend on t. This approach is employed next to design a distributed detection system.

Distributed Minimum Equivocation Detection

We consider the binary hypothesis testing problem for the distributed detection system in a parallel topology without the fusion center. The observations at the local detectors are again assumed to be conditionally

7.2 Distributed Detection Based on Information Theoretic Criterion 239

independent. The goal here is to find optimum decision rules at the individual detectors that maximize the mutual information $I(H; \mathbf{u})$ where the vector $\mathbf{u}$ consists of all the local decisions. It can easily be seen that the optimum local detectors in this case are threshold detectors. Therefore, it suffices to derive the expressions for optimum thresholds at individual detectors that maximize $I(H; \mathbf{u})$. The likelihood ratio test at each detector is given by

$$\Lambda(y_i) \begin{array}{c} u_i = 1 \\ > \\ < \\ u_i = 0 \end{array} t_i, \quad i = 1, \ldots, N, \qquad (7.2.11)$$

where the local threshold t_i is

$$t_i = \frac{\partial P_{Di}}{\partial P_{Fi}}, \quad i = 1, \ldots, N. \qquad (7.2.12)$$

Here, we consider the case of two detectors in detail, i.e., we consider the system shown in Figure 3.1. The two-detector case without fusion can be represented in terms of the equivalent channel model shown in Figure 7.3. The transition probabilities $P(u_1, u_2|H)$ are indicated in terms of P_{Fi} and P_{Di}, respectively. The output probabilities $P(\mathbf{u}^T = 00)$, $P(\mathbf{u}^T = 01)$, $P(\mathbf{u}^T = 10)$, and $P(\mathbf{u}^T = 11)$ are denoted by α_{00}, α_{01}, α_{10}, and α_{11}, respectively. They are given by

$$\alpha_{00} = P_0(1-P_{F1})(1-P_{F2}) + (1-P_0)(1-P_{D1})(1-P_{D2}),$$

$$\alpha_{01} = P_0(1-P_{F1})P_{F2} + (1-P_0)(1-P_{D1})P_{D2},$$

$$\alpha_{10} = P_0 P_{F1}(1-P_{F2}) + (1-P_0)P_{D1}(1-P_{D2}),$$

and

$$\alpha_{11} = P_0 P_{F1} P_{F2} + (1-P_0) P_{D1} P_{D2}.$$

Substituting the appropriate quantities, $I(H; \mathbf{u})$ may be expressed as

240   7. Information Theory and Distributed Hypothesis Testing

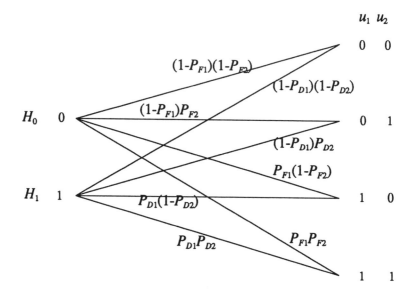

Figure 7.3. Channel model for the two-detector system without fusion.

$$\begin{aligned}
I(H; u) = &-P_0\{\log(P_0) + P_{F1}P_{F2}\log(P_0 P_{F1} P_{F2}) \\
&+ (1-P_{F1})(1-P_{F2})\log[P_0(1-P_{F1})(1-P_{F2})] \\
&+ (1-P_{F1})P_{F2}\log[P_0(1-P_{F1})P_{F2}] + P_{F1}(1-P_{F2})\log[P_0 P_{F1}(1-P_{F2})]\} \\
&+ (1-P_0)\{(1-P_{D1})(1-P_{D2})\log[(1-P_0)(1-P_{D1})(1-P_{D2})] \\
&+ (1-P_{D1})P_{D2}\log[(1-P_0)(1-P_{D1})P_{D2}] \\
&+ P_{D1}(1-P_{D2})\log[(1-P_0)P_{D1}(1-P_{D2})] \\
&+ P_{D1}P_{D2}\log[(1-P_0)P_{D1}P_{D2}] - \log(1-P_0)\} \\
&- [\alpha_{00}\log(\alpha_{00}) + \alpha_{01}\log(\alpha_{01}) + \alpha_{10}\log(\alpha_{10}) + \alpha_{11}\log(\alpha_{11})] .
\end{aligned}$$

(7.2.13)

Taking the derivative of $I(H; u)$ with respect to P_{F1},

$$\frac{\partial I(H; \mathbf{u})}{\partial P_{F1}} = -P_0 \log\left(\frac{1-P_{F1}}{P_{F1}}\right) - (1-P_0)\frac{\partial P_{D1}}{\partial P_{F1}}\log\left(\frac{1-P_{D1}}{P_{D1}}\right)$$
$$-\frac{\partial}{\partial P_{F1}}(\alpha_{00}\log\alpha_{00} + \alpha_{10}\log\alpha_{10} + \alpha_{01}\log\alpha_{01} + \alpha_{11}\log\alpha_{11}).$$

(7.2.14)

Rearranging (7.2.14), setting the result equal to zero, and solving for t_1,

$$t_1 = -\frac{P_0\left[\log\left(\frac{\alpha_{00}}{\alpha_{10}}\right) + P_{F2}\log\left(\frac{\alpha_{01}\alpha_{10}}{\alpha_{00}\alpha_{11}}\right) - \log\left(\frac{1-P_{F1}}{P_{F1}}\right)\right]}{(1-P_0)\left[\log\left(\frac{\alpha_{00}}{\alpha_{10}}\right) + P_{D2}\log\left(\frac{\alpha_{01}\alpha_{10}}{\alpha_{00}\alpha_{11}}\right) - \log\left(\frac{1-P_{D1}}{P_{D1}}\right)\right]}.$$

(7.2.15)

Similarly,

$$t_2 = -\frac{P_0\left[\log\left(\frac{\alpha_{00}}{\alpha_{10}}\right) + P_{F1}\log\left(\frac{\alpha_{01}\alpha_{10}}{\alpha_{00}\alpha_{11}}\right) - \log\left(\frac{1-P_{F2}}{P_{F2}}\right)\right]}{(1-P_0)\left[\log\left(\frac{\alpha_{00}}{\alpha_{10}}\right) + P_{D1}\log\left(\frac{\alpha_{01}\alpha_{10}}{\alpha_{00}\alpha_{11}}\right) - \log\left(\frac{1-P_{D2}}{P_{D2}}\right)\right]}.$$

(7.2.16)

As can be readily seen, t_1 and t_2 are coupled, and a simultaneous solution is required for their computation.

Next, we discuss the design of the parallel fusion network, shown in Figure 3.6, based on an information theoretic criterion. The fusion rule and the decision rules at the peripheral detectors are designed so that they maximize the mutual information $I(H;u_0)$. It can be shown that these decision rules can be implemented in terms of likelihood ratio tests [WaW89]. Once again, the person-by-person optimization methodology will be employed for system optimization. Define a logarithmic cost function as follows:

$$C_{ij} = \log\left(\frac{P(u_{0i}, H_j)}{P(u_{0i})P(H_j)}\right), \quad i, j = 0, 1, \quad (7.2.17)$$

where u_{0i} represents $u_0 = i$. Based on this cost, the mutual information $I(H; u_0)$ can be expressed as

$$I(H; u_0) = P_0(C_{10} - C_{00})P_F - P_1(C_{01} - C_{11})P_D + P_0 C_{00} + P_1 C_{01}.$$

(7.2.18)

Let v be a variable so that $v \in (a, b)$, and $I(H; u_0)$ is a function of v. The value v_{ext} of v, where the mutual information has an extremum (if the extremum exists and lies in (a, b)), is obtained by setting the derivative of $I(H; u_0)$, with respect to v, equal to zero. Taking the partial derivative of $I(H; u_0)$ with respect to v and rearranging the result,

$$\frac{\partial I(H; u_0)}{\partial v} = F(\partial v) + P_0(C_{10} - C_{00})\frac{\partial P_F}{\partial v} - P_1(C_{01} - C_{11})\frac{\partial P_D}{\partial v},$$

(7.2.19)

where

$$F(\partial v) = P_0\left(\frac{\partial C_{10}}{\partial v} - \frac{\partial C_{00}}{\partial v}\right)P_F - P_1\left(\frac{\partial C_{01}}{\partial v} - \frac{\partial C_{11}}{\partial v}\right)P_D$$
$$+ P_0\frac{\partial C_{00}}{\partial v} + P_1\frac{\partial C_{01}}{\partial v},$$

or, equivalently,

$$F(\partial v) = \frac{\partial C_{00}}{\partial v}[P_0(1 - P_F)] + \frac{\partial C_{01}}{\partial v}[P_1(1 - P_D)]$$
$$+ \frac{\partial C_{10}}{\partial v}P_0 P_F + \frac{\partial C_{11}}{\partial v}P_1 P_D.$$

After some manipulation, it can be shown that $F(\partial v)$ is equal to zero

[HoV89a] and $\partial I(H; u_0)/\partial v$ becomes

$$\frac{\partial I(H; u_0)}{\partial v} = P_1(C_{11}-C_{01})\frac{\partial P_D}{\partial v} - P_0(C_{00}-C_{10})\frac{\partial P_F}{\partial v}. \tag{7.2.20}$$

Whenever this expression is greater than (less than) zero, $I(H; u_0)$ is an increasing (decreasing) function in v. When this expression is zero, it can be solved for v that corresponds to an extremum.

The PBPO fusion rule is determined in terms of P_{1u^*}, the probability of deciding $u_0 = 1$ when the incoming decision vector is equal to u^*. In this case,

$$\frac{\partial I(H; u_0)}{\partial P_{1u^*}} = P_1(C_{11}-C_{01})\frac{\partial P_D}{\partial P_{1u^*}} - P_0(C_{00}-C_{10})\frac{\partial P_F}{\partial P_{1u^*}}. \tag{7.2.21}$$

Recall from Chapter 5 that

$$P_D = \sum_u P_{1u} M_u,$$

and

$$P_F = \sum_u P_{1u} F_u.$$

Substituting these in (7.2.21),

$$\frac{\partial I(H; u_0)}{\partial P_{1u^*}} = P_1(C_{11}-C_{01})M_{u^*} - P_0(C_{00}-C_{10})F_{u^*}. \tag{7.2.22}$$

The fusion rule is obtained by observing that, whenever (7.2.22) is positive (negative), $I(H; u_0)$ is an increasing (decreasing) function of P_{1u^*}. Therefore, P_{1u^*} should be set to its maximum (minimum) value of 1(0) when (7.2.22) is positive (negative). Thus, the fusion rule is

$$P_1(C_{11}-C_{01})M_{u^*} - P_0(C_{00}-C_{10})F_{u^*} \begin{matrix} \text{set}P_{1u^*}=1 \\ > \\ < \\ \text{set}P_{1u^*}=0 \end{matrix} 0. \quad (7.2.23)$$

We assume that $C_{00} \geq C_{10}$ and $C_{11} \geq C_{01}$, which is equivalent to the reasonable assumption

$$\left[\frac{P(u_0=i|H_i)}{P(u_0=i|H_j)}\right] \geq 1, \quad i \neq j, \ i, j = 0, 1. \quad (7.2.24)$$

Therefore, the fusion rule can be expressed as

$$\frac{M_{u^*}}{F_{u^*}} = \frac{P(u^*|H_1)}{P(u^*|H_0)} \begin{matrix} P(u_0=1|u^*)=1 \\ > \\ < \\ P(u_0=1|u^*)=0 \end{matrix} \frac{P_0(C_{00}-C_{10})}{P_1(C_{11}-C_{01})}. \quad (7.2.25)$$

The PBPO decision rule at DM k, $k = 1, ..., N$, can be obtained by setting the following derivative of $I(H; u_0)$, with respect to P_{Fk}, equal to zero:

$$\frac{\partial I(H; u_0)}{\partial P_{Fk}} = P_1(C_{11}-C_{01})\frac{\partial P_D}{\partial P_{Fk}} - P_0(C_{00}-C_{10})\frac{\partial P_F}{\partial P_{Fk}}. \quad (7.2.26)$$

Using the fact that the threshold t_k is given by

$$t_k = \frac{\partial P_{Dk}}{\partial P_{Fk}},$$

and the chain rule,

$$\frac{\partial P_D}{\partial P_{Fk}} = \frac{\partial P_D}{\partial P_{Dk}} t_k. \quad (7.2.27)$$

Substituting (7.2.27) in (7.2.26),

$$\frac{\partial I(H; u_0)}{\partial P_{Fk}} = P_1(C_{11} - C_{01})\frac{\partial P_D}{\partial P_{Dk}} t_k - P_0(C_{00} - C_{10})\frac{\partial P_F}{\partial P_{Fk}}. \qquad (7.2.28)$$

Setting (7.2.28) equal to zero and solving for t_k,

$$t_k = \frac{P_0(C_{10} - C_{00})\dfrac{\partial P_F}{\partial P_{Fk}}}{P_1(C_{01} - C_{11})\dfrac{\partial P_D}{\partial P_{Dk}}}, \quad k=1, \ldots, N. \qquad (7.2.29)$$

As expected, the expression for t_k obtained here has the same form as the one found in the Bayesian case, except that, here, C_{ij}, $i, j = 0, 1$ are no longer constants. A simultaneous solution of (7.2.25) and (7.2.29) yields the desired solution.

Design of distributed detection systems based on an information theoretic criterion provides an alternate framework for system optimization. This approach is attractive for system design in certain situations. It also provides an intuitively appealing view of the distributed detection problem as an information transmission problem.

7.3 Multiterminal Detection with Data Compression

Consider the conventional binary hypothesis testing problem using a single sensor. The observations are generated by a discrete-time, finite-alphabet, memoryless source. The two hypotheses are characterized by their density functions H_0: $p(y_i)$ and H_1: $q(y_i)$, $y_i \in Y$, where Y is the source alphabet. On the basis of n observations $(y_1, \ldots, y_n) \in Y^n$, the detector is to determine which hypothesis is true under the Neyman–Pearson criterion. Let $A_n \subset Y^n$ be an acceptance region for hypothesis H_0 on the basis of n observations, i.e., for all $(y_1, \ldots, y_n) \in A_n$, hypothesis H_0 is decided. The complement of A_n over Y^n is denoted by $\overline{A}_n$. The probability of false alarm α_n and the probability of miss β_n are defined as

$$\alpha_n = \text{Prob}\left[(y_1, ..., y_n) \in \bar{A}_n | H_0\right]$$

$$= p^n(\bar{A}_n), \qquad (7.3.1)$$

and

$$\beta_n = \text{Prob}\left[(y_1, ..., y_n) \in A_n | H_1\right]$$

$$= q^n(A_n), \qquad (7.3.2)$$

where $p^n(\cdot)$ and $q^n(\cdot)$ are probability measures on $\mathbf{Y}^n$. Let $\varepsilon \in (0,1)$ be the specified constraint on the probability of false alarm, i.e., the test must be designed so that $\alpha_n \leq \varepsilon$. Define the minimum probability of miss to be

$$\beta_n(\varepsilon) = \min_{\substack{A_n \subset \mathbf{Y}^n \\ \alpha_n \leq \varepsilon}} \beta_n. \qquad (7.3.3)$$

The minimum probability of miss $\beta_n(\varepsilon)$ converges to zero exponentially as n goes to infinity. The rate of convergence is provided by Stein's lemma [Che52, CoT91] given as follows:

$$\lim_{n \to \infty} \frac{1}{n} \log \beta_n(\varepsilon) = -D(p \| q), \qquad (7.3.4)$$

where $D(p \| q)$ is the relative entropy or the Kullback–Leibler informational divergence between distributions p and q, given as

$$D(p \| q) = \sum_{y \in \mathbf{Y}} p(y) \log \frac{p(y)}{q(y)}. \qquad (7.3.5)$$

If the constraint on P_F is in the form of an exponential, i.e., $\alpha_n \leq e^{-nr}$, $r>0$, then, $\beta_n \approx e^{-ne(r)}$ where $e(r)$ is the error exponent function [Bla74].

A number of attempts have been made to extend the above results to distributed detection problems. Several important results have been obtained. Consider a distributed detection system consisting of two local detectors DM x and DM y and a fusion center. Observations at the two local detectors are denoted by x_i, y_i, $i = 1, ..., n$. These observations are generated by a discrete-time, discrete-alphabet, memoryless, multiple source. The detectors DM x and DM y encode their observations into a maximum of M_n and N_n messages, respectively. The encoded messages are sent to the fusion center which makes a decision as to which hypothesis is present. In most practical situations, fixed codebook sizes are used, i.e., $M_n = M$ and $N_n = N$. For the hard decision system, $M = 2$ and $N = 2$. When M_n and N_n are large enough, no compression is needed, and the fusion center has access to the observations x_i, y_i, $i = 1, ..., n$. In this case, the problem reduces to a centralized detection problem, and the result (7.3.4), based on Stein's lemma, is still valid. However, when data compression is mandatory, i.e., there are communication constraints on channels linking the local detectors and the fusion center, the problem becomes more challenging. Nonetheless, several important contributions have been made that are briefly discussed next.

Ahlswede and Csiszar [AhC86] considered the two-sensor distributed detection system in which the fusion center receives complete observations (without compression) from one of the sensors and data compressed at a specified rate R from the other sensor. Testing the hypothesis of a given bivariate distribution $p(x, y)$ against the alternative of independence was considered, i.e., H_0: $p(x_i, y_i)$ and H_1: $p(x_i) p(y_i)$, $x_i \in X$, $y_i \in Y$. Observations y_i, $i = 1, ..., n$, from DM y are directly available at the fusion center, but those from DM x are not. Instead, observations $X_n = \{x_1, ..., x_n\}$ are encoded at DM x via an encoding function of rate R and the encoded version $\gamma_x(X_n)$ is made available at the fusion center. Rate constraint is given by

$$\frac{1}{n} \log |\gamma_x| \leq R ,\qquad(7.3.6)$$

where $|\gamma_x|$ represents the range of the encoding function γ_x. For any fixed rate R, $R \geq 0$, and probability of false alarm less than or equal to ε, $\varepsilon \in (0,1)$, the asymptotic behavior of the smallest achievable probability of miss is given by

$$\lim_{n \to \infty} \frac{1}{n} \log \beta_n(\varepsilon, R) = -\theta(R) , \qquad (7.3.7)$$

where

$$\theta(R) = \sup_k \theta_k(R) ,$$

and

$$\theta_k(R) = \sup_{\gamma_x} \left\{ \frac{1}{k} D[p(\gamma_x(X_k), Y_k) \| p(\gamma_x(X_k)) p(Y_k)] \big| \log |\gamma_x| \leq kR \right\}.$$

The more general problem of bivariate hypotheses with one-sided data compression was also considered. Let (x_i, y_i), $i=1, ..., n$, denote the observations at the two sensors. Let $p(x_i, y_i)$ and $q(x_i, y_i)$ denote the two bivariate distributions under the two hypotheses, so that the hypothesis testing problem is given as H_0: $p(x_i, y_i)$ and H_1: $q(x_i, y_i)$. In this one-sided data compression problem, the fusion center receives direct observations from DM y and encoded messages via an encoding function of rate R from DM x. Assuming that $q(x_i, y_i) > 0$, $\forall\, x_i \in X$ and $y_i \in Y$, the asymptotic value of smallest achievable probability of miss for all $R \geq 0$ and $\varepsilon \in (0,1)$ is again given by (7.3.7), where

$$\theta(R) = \sup_k \theta_k(R) ,$$

and

$$\theta_k(R) = \sup_{\gamma_x} \left\{ \frac{1}{k} D[p(\gamma_x(X_k), Y_k) \| q(\gamma_x(X_k), Y_k)] \big| \log |\gamma_x| \leq kR \right\}.$$

The exponential rate at which the probability of miss goes to zero as $n \to \infty$ is independent of ε. The lower bound on $\theta(R)$ was also derived. This problem was further investigated by Han [Han87]. He considered the general case in which both local detectors encoded their observations at rates R_1 and R_2 prior to transmission to the fusion center. He obtained two lower bounds on the power exponent function and derived a number of interesting properties. His bound was tighter than that of Ahlswede and Csiszar [AhC86].

Han and Kobayashi [HaK89] studied the multiterminal hypothesis testing problem H_0: $p(x_i, y_i)$ and H_1: $q(x_i, y_i)$ and considered the case in which the constraint on the probability of false alarm is assumed to be of the exponential type, i.e., $\alpha_n \leq e^{-nr}$, $r > 0$ instead of the type $\alpha_n < \varepsilon$. The general case, where both local detectors compress their observations at fixed rates R_1 and R_2, as well as the special case, where one local detector sends uncompressed observations and the other one compresses at rate R are studied. Lower bounds on the exponent were established. Shalaby and Papamarcou [ShP92] considered the multiterminal hypothesis testing problem under data compression at (asymptotically) zero rate. They assumed that the codebook sizes satisfied constraints of the type

$$R_1(n) = \frac{1}{n}\log M_n \to 0 \quad \text{as } n \to \infty,$$

and

$$R_2(n) = \frac{1}{n}\log N_n \to 0 \quad \text{as } n \to \infty.$$

Asymptotic optimality of the distributed detection system was investigated. For a simple hypothesis testing problem under the positivity constraint $q(x_i, y_i) > 0$ and for a fixed value of ε, it was shown that one local detector transmitting data to the fusion center at a vanishing rate while the other supplies complete information about its observations is asymptotically no better than the optimal system in which each local detector transmits a single binary digit to the fusion center. Another interesting result for the simple hypothesis testing problem was that, for fixed codebook-size compression, the codebook sizes M and N at the two detectors do not affect asymptotic performance. Thus, transmission of soft decisions from the local detectors, instead of hard decisions, does not improve system performance in the asymptotic sense. Unlike the simple hypothesis testing problem, the error exponent for the composite hypothesis testing problem is a function of the two distribution classes, the probability of false alarm ε and the codebook sizes M_n and N_n.

Notes and Suggested Reading

Design of detection systems based on information-theoretic cost functions was discussed in[Mid60]. Further results are available in[Gab66], [Mar86], and [HoV89a]. This approach has been applied to the design of distributed detection systems in [HoV89a]. Details on asymptotic performance of multiterminal detection systems are available in [AhC86], [Han87], [HaK89], [AmH89], [ShP92], and [ChP93]. Other related results are given in [Tsi88] and [Kaz91].

Selected Bibliography

[Aal91] Aalo, V., "Performance Study of Some Distributed Detection Rules," Ph.D. Dissertation, Southern Illinois University, Aug. 1991.

[AaV89] Aalo, V., and R. Viswanathan, "On Distributed Detection with Correlated Sensors: Two Examples," *IEEE Trans. on Aerospace and Elect. Syst.*, Vol. 25, No. 3, pp. 414–421, May 1989.

[AaV92] Aalo, V., and R. Viswanathan, "Asymptotic Performance of a Distributed Detection System in Correlated Gaussian Noise," *IEEE Trans. on Signal Proc.*, Vol. 40, No. 1, pp. 211–213, Jan. 1992.

[AhC86] Ahlswede, R., and I. Csiszar, "Hypothesis Testing with Communication Constraints," *IEEE Trans. on Info. Theory*, IT-32, No. 4, pp. 533–542, July 1986.

[AHZ96] Ansari, N., E.S.H. Hou, B. Zhu, and J. Chen, "Adaptive Fusion by Reinforcement Learning for Distributed Detection Systems," *IEEE Trans. on Aerospace and Elect. Syst.*, Vol. 32, No. 2, pp. 524–531, April 1996.

[AlB87] Al–Bassiouni, A.M., "Optimum Signal Processing in Distributed Sensor Systems," Ph.D. Dissertation, Naval Postgraduate School, Dec. 1987.

Selected Bibliography

[Alh90]   Alhakeem, S., "Decentralized Bayesian Hypothesis Testing with Feedback," Ph.D. Dissertation, Syracuse University, Dec. 1990.

[AlI89]   Al–Ibrahim, M.M., "On Distributed Sequential Hypothesis Testing," Ph.D. Dissertation, Syracuse University, Dec. 1989.

[AlV88]   Al–Ibrahim, M.M., and P.K. Varshney, "A Simple Multi-Sensor Sequential Detection Procedure," *Proc. 27th IEEE Conf. on Decision and Control,* Austin, Texas, 1988.

[AlV89]   Al–Ibrahim, M.M., and P.K. Varshney, "A Decentralized Sequential Test with Data Fusion," *Proc. 1989 American Control Conf.,* Pittsburgh, Pennsylvania, 1989.

[AlV90]   Alhakeem, S., and P.K. Varshney, "A Bayesian Formulation of Decentralized Detection Systems with Feedback," *Proc. 1990 Conf. on Info. Sciences and Systems,* Princeton, March 1990.

[AlV95]   Alhakeem, S., and P.K. Varshney, "A Unified Approach to the Design of Decentralized Detection Systems," *IEEE Trans. on Aerospace and Elect. Syst.,* Vol. 31, No. 1, pp. 9–20, Jan. 1995.

[AlV96]   Alhakeem, S., and P.K. Varshney, "Decentralized Bayesian Detection with Feedback," *IEEE Trans. on Systems, Man and Cybernetics,* Vol. 26, No. 4, pp. 503–513, July 1996.

[AmH89]   Amari, S., and T.S. Han, "Statistical Inference under Multiterminal Rate Restrictions: A Differential Geometric Approach," *IEEE Trans. on Info. Theory,* Vol. 35, No. 2, pp. 217–227, March 1989.

[Bar87]   Baras, J.S., "Distributed Asynchronous Detection: General Models," *Proc. 26th IEEE Conf. on Decision and Control,* pp. 1832–1835, 1987.

[Bar91]   Barkat, M., *Signal Detection and Estimation,* Artech House, Boston, 1991.

[BaV89]   Barkat, M., and P.K. Varshney, "Decentralized CFAR Signal Detection," *IEEE Trans. on Aerospace and Elect. Syst.*, Vol. 25, No. 2, pp. 141–149, March 1989.

[BaV91]   Barkat, M., and P.K. Varshney, "Adaptive Cell-Averaging CFAR Detection in Distributed Sensor Networks," *IEEE Trans. on Aerospace and Elect. Syst.*, Vol. 27, No. 3, pp. 424–429, May 1991.

[Ber82]   Bertsekas, D., *Constrained Optimization and Lagrange Multiplier Methods*, Academic Press, New York, 1982.

[BKP93]   Batalama, S.N., A.G. Koyiantis, P. Papantoni–Kazakos, and D. Kazakos, "Feedforward Neural Structures in Binary Hypothesis Testing," *IEEE Trans. on Communications*, Vol. COM-41, No. 6, pp. 1047–1062, July 1993.

[Bla74]   Blahut, R.E., "Hypothesis Testing and Information Theory," *IEEE Trans. on Info. Theory*, Vol. IT-20, No. 4, pp. 405–417, July 1974.

[BlK92]   Blum, R.S., and S.A. Kassam, "Optimum Distributed Detection of Weak Signals in Dependent Sensors," *IEEE Trans. on Info. Theory*, Vol. 38, No. 3, pp. 1066–1079, May 1992.

[BlK95a]   Blum, R.S., and S.A. Kassam," Distributed Cell-Averaging CFAR Detection in Dependent Sensors," *IEEE Trans. on Info. Theory*, Vol. 41, No. 2, pp. 513–518, March 1995.

[BlK95b]   Blum, R.S., and S.A. Kassam, "On the Asymptotic Relative Efficiency of Distributed Detection Schemes," *IEEE Trans. on Info. Theory*, Vol. 41, No. 2, pp. 523–527, March 1995.

[BlK95c]   Blum, R.S., and S.A. Kassam, "Optimum Distributed CFAR Detection of Weak Signals," *Journal of the Acoustical Society of America*, Vol. 98, No. 1, pp. 221–229, July 1995.

[BlQ96]   Blum, R.S., and J. Qiao, "Threshold Optimization for Distributed Order-Statistic CFAR Signal Detection," *IEEE Trans. on Aerospace and Elect. Syst.*, Vol. 32, pp. 568–577, Jan. 1996.

[Blu95a] Blum, R.S., "Quantization in Multisensor Random Signal Detection," *IEEE Trans. on Info. Theory*, Vol. 41, No. 1, pp. 204–215, Jan. 1995.

[Blu95b] Blum, R.S., "Distributed Detection of Narrowband Signals," *IEEE Trans. on Info. Theory*, Vol. 41, No. 2, pp. 519–523, March 1995.

[Blu96a] Blum, R.S., "Locally Optimum Distributed Detection of Dependent Random Signals Based on Ranks," *IEEE Trans. on Info. Theory*, Vol. 42, No. 3, pp. 931–942, May 1996.

[Blu96b] Blum, R.S., "Necessary Conditions for Optimum Distributed Sensor Detectors under the Neyman–Pearson Criterion," *IEEE Trans. on Info. Theory*, Vol. 42, No. 3, pp. 990–994, May 1996.

[Bog87] Bogler, P.L., "Shafer–Dempster Reasoning with Applications to Multisensor Target Identification Systems," *IEEE Trans. on Systems, Man, and Cybernetics*, Vol. SMC-17, No. 6, pp. 968–977, Nov./Dec. 1987.

[BoT86] Boettcher, K.L., and R.R. Tenney, "Distributed Decisionmaking with Constrained Decisionmakers," *IEEE Trans. on Systems, Man, and Cybernetics,* Vol. SMC-16, No. 6, pp. 813–823, Nov./Dec. 1986.

[CDF83] Conte, E., E. D'Addio, A. Farina, and M. Longo, "Multistatic Radar Detection: Synthesis and Comparison of Optimum and Suboptimum Receivers," *IEE Proc.,* 130, Part F, pp. 484–494, 1983.

[CDL87] Chao, J.J., E. Drakopoulos, and C.C. Lee, "An Evidential Reasoning Approach to Distributed Multiple-Hypothesis Testing," *Proc. 26th IEEE Conf. on Decision and Control*, pp. 1826–1831, 1987.

[Che52] Chernoff, H., "A Measure of Asymptotic Efficiency for Tests of a Hypothesis Based on a Sum of Observations," *Annals of Mathematical Statistics*, Vol. 23, pp. 493–507, 1952.

[ChK92]   Cherikh, M., and P.B. Kantor, "Counterexamples in Distributed Detection," *IEEE Trans. on Info. Theory*, Vol. 38, No. 1, pp. 162–165, Jan. 1992.

[ChK94]   Chang, W., and M. Kam, "Asynchronous Distributed Detection," *IEEE Trans. on Aerospace and Elect. Syst.*, Vol. 30, pp. 818–826, July 1994.

[ChP93]   Chen, P., and A. Papamarcou, "New Asymptotic Results in Parallel Distributed Detection," *IEEE Trans. on Info. Theory*, Vol 39, No. 6, pp. 1847–1863, Nov. 1993.

[ChP95]   Chen, P., and A. Papamarcou, "Error Bounds for Parallel Distributed Detection under the Neyman–Pearson Criterion," *IEEE Trans. on Info. Theory*, Vol. 41, pp. 528–533, March 1995.

[ChV86]   Chair, Z., and P.K. Varshney, "Optimal Data Fusion in Multiple Sensor Detection Systems," *IEEE Trans. on Aerospace and Elect. Syst.*, Vol. AES-22, No. 1, pp. 98–101, Jan. 1986.

[ChV87]   Chair, Z., and P.K. Varshney, "Neyman–Pearson Hypothesis Testing in Distributed Networks," *Proc. 26th IEEE Conf. on Decision and Control*, pp. 1842–1843, 1987.

[ChV88]   Chair, Z., and P.K. Varshney, "Distributed Bayesian Hypothesis Testing with Distributed Data Fusion," *IEEE Trans. on Systems, Man, and Cybernetics*, Vol. SMC-18, No. 5, pp. 695–699, Sept./Oct. 1988.

[CoT91]   Cover, T.M., and J.A. Thomas, *Elements of Information Theory*, John Wiley & Sons, New York, 1991.

[CrS96]   Crow, R.W., and S.C. Schwartz, "Quickest Detection for Sequential Decentralized Decision Systems," *IEEE Trans. on Aerospace and Elect. Systems*, Vol. 32, No. 1, pp. 267–283, Jan. 1996.

[CsL71]   Csiszar, I., and G. Longo, "On the Error Exponent for Source Coding and for Testing Simple Statistical

Hypotheses," *Studia Sci. Math. Hungar.*, Vol. 6, pp. 181–191, 1971.

[Das91] Dasarathy, B.V., "Decision Fusion Strategies in Multisensor Environments," *IEEE Trans. on Systems, Man, and Cybernetics*, Vol. 21, No. 5, pp. 1140–1154, Sept./Oct. 1991.

[Das94] Dasarathy, B.V., *Decision Fusion*, IEEE Computer Society Press, Los Alamitos, CA, 1994.

[Dav81] David, H.A., *Order Statistics*, John Wiley & Sons, New York, 1981.

[DCL92] Drakopoulos, E., J.J. Chao, and C.C. Lee, "A Two-Level Distributed Multiple Hypothesis Decision System," *IEEE Trans. on Automatic Control*, Vol. 37, No. 3, pp. 380–384, March 1992.

[Dem88] Demirbas, K., "Maximum A Posteriori Approach to Object Recognition with Distributed Sensors," *IEEE Trans. on Aerospace and Elect. Syst.*, Vol. AES-24, No. 3, pp. 309–313, May 1988.

[Dem89] Demirbas, K., "Distributed Sensor Data Fusion with Binary Decision Trees," *IEEE Trans. on Aerospace and Elect. Syst.*, Vol. AES-25, No. 5, pp. 643–649, Sept. 1989.

[DeP93] Delic, H., and P. Papantoni-Kazakos, "Robust Decentralized Detection by Asympototically Many Sensors," *Signal Processing*, Vol. 33, No. 2, pp. 223–234, Aug. 1993.

[DoB91] Donohue, K.D., and N.M. Bilgutay, "OS Characterization for Local CFAR Detection," *IEEE Trans. on Systems, Man, and Cybernetics*, Vol. 21, No. 5, pp. 1212–1216, Sept./Oct. 1991.

[Dom87] Dommermuth, F.M., "Distributed Radar Detection Theory On the Most Powerful Decentralized Test," *IEE Proc.*, Vol. 134, pp. 202–204, 1987.

[DPK95] Delic, H., P. Papantoni-Kazakos, and D. Kazakos, "Fundamental Structures and Asymptotic Performance Criteria in Decentralized Binary Hypothesis Testing," *IEEE Trans. on Commun.*, Vol. 43, No. 1, pp. 32–43, Jan. 1995.

[DrL91] Drakopoulos, E., and C.C. Lee, "Optimum Multisensor Fusion of Correlated Local Decisions," *IEEE Trans. on Aerospace and Elect. Syst.*, Vol. AES-27, No. 4, July 1991, pp. 593–605.

[DrL92] Drakopoulos, E., and C.C. Lee, "Decision Rules for Distributed Decision Networks with Uncertainties," *IEEE Trans. on Automatic Control*, Vol. 37, No. 1, pp. 5–14, Jan. 1992.

[DuH73] Duda, P.O., and P.E. Hart, *Pattern Classification and Scene Analysis*, John Wiley & Sons, New York, 1973.

[DuM88] Dudewicz, E.J., and S.N. Mishra, *Modern Mathematical Statistics*, John Wiley & Sons, New York, 1988.

[EBA92] Elias–Fusté, A.R., A. Broquetas–Ibars, J.P. Antequera, and J.C.M. Yuste, "CFAR Data Fusion Center with Inhomogeneous Receivers," *IEEE Trans. on Aerospace and Elect. Syst.*, Vol. 28, No. 1, pp. 276–285, Jan. 1992.

[Ekc93] Ekchian, L.K., "Optimal Design of Distributed Detection Networks," Ph.D. Thesis, Dept. of Electrical Engineering and Computer Science, M.I.T., Cambridge, Massachusetts, 1983.

[EkT82] Ekchian, L.K., and R.R. Tenney, "Detection Networks," *Proc. 21st IEEE Conf. on Decision and Control*, pp. 686–691, 1982.

[EkT83] Ekchian, L.K., and R.R. Tenney, "Recursive Solution of Distributed Detection/Communication Problems," *Proc. 1983 American Controls Conference*, pp. 1338–1339, Vol. 3, 1983.

[FiJ68]   Finn, H.M., and R.S. Johnson, "Adaptive Detection Mode with Threshold Control as a Function of Spatially Sample Clutter-level Estimates," *RCA Review*, Vol. 29, pp. 414–464, Sept. 1968.

[FlG87]   Flynn, T.J., and R.M. Gray, "Encoding of Correlated Observations," *IEEE Trans. on Info. Theory*, Vol. IT-33, pp. 773–787, Nov. 1987.

[Fly85]   Flynn, T.J., "Quantizer Design for Distributed Sensing," Ph.D. Dissertation, Stanford University, March 1985.

[Gab66]   Gabrielle, T.L., "Information Criterion for Threshold Determination," *IEEE Trans. on Info. Theory*, Vol. 6, pp. 484–486, Oct. 1966.

[GaK88]   Gandhi, P.P., and S.A. Kassam, "Analysis of CFAR Processor in Nonhomogeneous Background," *IEEE Trans. on Aerospace and Elect. Syst.*, Vol. 24, pp. 427–445, July 1988.

[GaJ79]   Garey, M.R., and D.S. Johnson, *Computers and Intractability: A Guide to the Theory of NP-Completeness*, W.H. Freeman, San Francisco, 1979.

[GeC90]   Geraniotis, E., and Y.A. Chau, "Robust Data Fusion for Multisensor Detection Systems," *IEEE Trans. on Info. Theory*, Vol. 36, No. 6, pp. 1265–1279, Nov. 1990.

[GoV95]   Gowda, C.H., and R. Viswanathan, "Robustness of Decentralized Tests with ε-Contamination Prior," *IEEE Trans. on Info. Theory*, Vol. 41, No. 4, pp. 1164–1169, July 1995.

[HAE95]   Hussaini, E.K, A.A.M. Al–Bassiouni, and Y.A. El–Far, "Decentralized CFAR Signal Detection," *Signal Processing*, Vol. 44, pp. 299–307, July 1995.

[HaK89]   Han, T.S., and K. Kobayashi, "Exponential-type Error Probabilities for Multiterminal Hypothesis Testing," *IEEE Trans. on Info. Theory*, Vol. 35, pp. 2–14, Jan. 1989.

[Han87] Han, T.S., "Hypothesis Testing with Multiterminal Data Compression," *IEEE Trans. on Info. Theory*, Vol. IT-33, pp. 759–772, Nov. 1987.

[Han92] Han, J., "New Results in Distributed Detection," Ph.D. Dissertation, Syracuse University, May 1992.

[HaR89] Hashemi, H.R., and I.B. Rhodes, "Decentralized Sequential Detection," *IEEE Trans. on Info. Theory*, Vol. IT-35, pp. 509–520, May 1989.

[Has91] Hashlamoun, W.A., "Applications of Distance Measures and Probability of Error Bounds to Distributed Detection Systems," Ph.D. Dissertation, Syracuse University, May 1991.

[HaV91] Hashlamoun, W.A., and P.K. Varshney, "An Approach to the Design of Distributed Bayesian Detection Structures," *IEEE Trans. on Systems, Man, and Cybernetics*, Vol. 21, No. 5, pp. 1206–1211, Sept./Oct. 1991.

[HaV93] Hashlamoun, W.A., and P.K. Varshney, "Further Results on Distributed Bayesian Signal Detection," *IEEE Trans. on Info. Theory*, Vol. 39, No. 5, pp. 1660–1661, Sept. 1993.

[Hel65] Helstrom, C.W., "The Performance of Sensors Connected in Parallel and in Coincidence," *IEEE Trans. on Commun. Technology*, pp. 191–195, June 1965.

[Hel95a] Helstrom, C.W., *Elements of Signal Detection and Estimation*, Prentice-Hall, Englewood Cliffs, N.J., 1995.

[Hel95b] Helstrom, C.W., "Gradient Algorithm for Quantization Levels in Distributed Detection Systems," *IEEE Trans. on Aerospace and Elect. Syst.*, Vol. 31, pp. 390–399, Jan. 1995.

[Hes75] Hestenes, M.R., *Optimization Theory The Finite Dimensional Case*, John Wiley & Sons, New York, 1975.

[Hob86] Hoballah, I.Y., "On the Design and Optimization of Distributed Signal Detection and Parameter Estimation

Systems," Ph.D. Dissertation, Syracuse University, Nov. 1986.

[HoV86]   Hoballah, I.Y., and P.K. Varshney, "Neyman–Pearson Detection with Distributed Sensors," *Proc. 25th IEEE Conf. on Decision and Control,* Athens, Greece, December 1986, pp. 237–241.

[HoV89a] Hoballah, I.Y., and P.K. Varshney, "An Information Theoretic Approach to the Distributed Detection Problem," *IEEE Trans. on Info. Theory,* Vol. IT-35, pp. 988–994, Sept. 1989.

[HoV89b] Hoballah, I.Y., and P.K. Varshney, "Distributed Bayesian Signal Detection," *IEEE Trans. on Info. Theory,* Vol. IT-35, pp. 995–1000, Sept. 1989.

[Hus94]   Hussain, A.M., "Multisensor Distributed Sequential Detection," *IEEE Trans. on Aerospace and Elec. Syst.,* Vol. 30, No. 3, pp. 698–708, July 1994.

[HVV90]   Han, J., P.K. Varshney, and V.C. Vannicola, "Some Results on Distributed Nonparametric Detection," *Proc. 29th IEEE Conf. on Decision and Control,* Honolulu, Dec. 1990.

[IKM91]   Iyengar, S.S., R.L. Kashyap, and R.N. Madan, "Distributed Sensor Networks — Introduction to the Special Section," *IEEE Trans. on Systems, Man, and Cybernetics,* Vol. 21, No.5, pp. 1027–1031, Sept./Oct. 1991.

[IrT94]   Irving, W.W., and J.N. Tsitsiklis, "Some Properties of Optimal Thresholds in Decentralized Detection," *IEEE Trans. on Automatic Control,* Vol. 39, No. 4, pp. 835–838, April 1994.

[JIK91]   Jayasimha, D.N., S.S. Iyengar, and R.L. Kashyap, "Information Integration and Synchronization in Distributed Sensor Networks," *IEEE Trans. on Systems, Man, and Cybernetics,* Vol. 21, No. 5, pp. 1032–1043, Sept./Oct. 1991.

[Kad94]   Kadota, T.T., "Integration of Complementary Detection-Localization Systems — An Example," *IEEE Trans. on Info. Theory*, Vol. 40, No. 3, pp. 808–819, May 1994.

[Kai67]   Kailath, T., "The Divergence and Bhattacharyya Distance Measure in Signal Selection," *IEEE Trans. on Commun.*, Vol. COM-15, No. 1, pp. 52–60, Feb. 1967.

[Kas88]   Kassam, S.A., *Signal Detection in Non-Gaussian Noise*, Springer–Verlag, New York, 1988.

[Kaz91]   Kazakos, D., "Asymptotic Error Probability Expressions for Multihypothesis Testing Using Multisensor Data," *IEEE Trans. on Systems, Man, and Cybernetics*, Vol. SMC-21, No. 5, pp. 1101–1114, Sept./Oct. 1991.

[Kaz92]   Kazakos, D., "Error Bounds and Optimum Quantization for Multisensor Distributed Signal Detection," *IEEE Trans. on Commun.*, Vol. 40, No. 7, pp. 1144–1151, July 1992.

[KCZ91]   Kam, M., W. Chang, and Q. Zhu, "Hardware Complexity of Binary Distributed Detection Systems with Isolated Local Bayesian Detectors," *IEEE Trans. on Systems, Man, and Cybernetics*, Vol. SMC-21, No.3, pp. 565–571, May/June 1991.

[Kle93]   Klein, L.A., "A Boolean Algebra Approach to Multiple Sensor Voting Fusion," *IEEE Trans. on Aerospace and Elect. Syst.*, Vol. 29, No. 2, pp. 317–327, April 1993.

[KrL90]   Krzysztofowicz, R., and D. Long, "Fusion of Detection Probabilities and Comparison of Multisensor Systems," *IEEE Trans. on Systems, Man, and Cybernetics*, Vol. SMC-20, pp. 665–677, May/June 1990.

[Kul59]   Kullback, S., *Information Theory and Statistics*, John Wiley & Sons, New York, 1959.

[KuP82]   Kushner, H.J., and A. Pacut, "A Simulation Study of a Decentralized Detection Problem," *IEEE Trans. on*

Automatic Control, Vol. AC-27, pp. 1116–1119, October 1982.

[KZG92] Kam, M., Q. Zhu, and W.S. Gray, "Optimal Data Fusion of Correlated Local Decisions in Multiple Sensor Detection Systems," *IEEE Trans. on Aerospace and Elect. Syst.*, Vol. 28, pp. 916–920, July 1992.

[LaS82] Lauer, G.S., and N.R. Sandell, Jr, "Distributed Detection with Waveform Observations: Correlated Observation Processes," *Proc. 1982 American Controls Conf.*, pp. 812–819, Vol.2, 1982.

[LeC89] Lee, C.C., and J.J. Chao, "Optimum Local Decision Space Partitioning for Distributed Detection," *IEEE Trans. on Aerospace and Elect. Syst.*, Vol. AES-25, pp. 536–544, July 1989.

[Leh86] Lehmann, E.L., *Testing Statistical Hypotheses*, John Wiley & Sons, New York, 1986.

[LiS93] Li, T., and I.K. Sethi, "Optimal Multilevel Decision Fusion with Distributed Sensors," *IEEE Trans. on Aerospace and Elect. Syst.*, Vol. 29, No. 4, pp. 1252–1259, Oct. 1993.

[LiS96] Li, T., and I.K. Sethi, "Distributed Decision Fusion in the Presence of Link Failures," *IEEE Trans. on Aerospace and Elect. Syst.*, Vol. 32, No. 2, pp. 661–667, April 1996.

[Llo92] Lloyd, S.P., "Least Squares Quantization in PCM," *IEEE Trans. on Info. Theory*, Vol. 28, No. 2, pp. 129–136, March 1982.

[LLG90] Longo, M., T.D. Lookabaugh, and R.M. Gray, "Quantization for Decentralized Hypothesis Testing under Communication Constraints," *IEEE Trans. on Info. Theory*, Vol. 36, No. 2, pp. 241–255, March 1990.

[LuK89]   Luo, R.C., and M.G. Kay, "Multisensor Integration and Fusion in Intelligent Systems," *IEEE Trans. on Systems, Man, and Cybernetics*, Vol. 19, No. 5, pp. 901–931, Sept./Oct. 1989.

[MaR72]   Marschak, J.R., and R. Radner, *The Economic Theory of Teams*, Yale University Press, New Haven, Connecticut, 1972.

[Mar86]   Martinez, A., "Detector Optimization Using Shannon's Information," *Proc. 20th Conf. on Info. Sciences and Systems*, Princeton, March 1986.

[MeC78]   Melsa, J.L., and D.L. Cohn, *Decision and Estimation Theory*, McGraw-Hill, New York, 1978.

[Mid60]   Middleton, D., *Statistical Communication Theory*, McGraw-Hill, New York, 1960.

[Nas93]   Nasipuri, A., "On Some Centralized and Distributed Parametric and Nonparametric Detection Schemes," Ph.D. Dissertation, University of Massachusetts, Sept. 1993.

[NaT93]   Nasipuri, A., and S. Tantaratana, "Nonparametric Distributed Detection Using Wilcoxon Statistics," *Proc. 1993 Conf. on Info. Sciences and Systems*, John Hopkins University, Baltimore, March 1993.

[PaA86]   Papastavrou, J.D., and M. Athans, "A Distributed Hypothesis-Testing Team Decision Problem with Communications Cost," *Proc. 25th IEEE Conf. on Decision and Control*, Athens, Greece, pp. 219–225, December 1986.

[PaA92a] Papastavrou, J.D., and M. Athans, "Distributed Detection by a Large Team of Sensors in Tandem," *IEEE Trans. on Aerospace and Elect. Syst.*, Vol. AES-28, No. 3, pp. 639–653, July 1992.

[PaA92b] Papastavrou, J.D., and M. Athans, "On Optimal Distributed Decision Architectures in a Hypothesis Testing

Environment," *IEEE Trans. on Automatic Control*, Vol. 37, No. 8, pp. 1154–1169, Aug. 1992.

[PaA95] Papastavrou, J.D., and M. Athans, "The Team ROC Curve in a Binary Hypothesis Testing Environment," *IEEE Trans. on Aerospace and Elect. Syst.*, Vol. 31, No. 1, pp. 96–105, Jan. 1995.

[Pap90] Papastavrou, J.D., "Decentralized Decision Making in a Hypothesis Testing Environment," Ph.D. Dissertation, M.I.T., May 1990.

[PaT86] Papadimitriou, C.H., and J.N. Tsitsiklis, "Intractable Problems in Control Theory," *SIAM Journal Control Optimiz.*, Vol. 24, pp. 639–654, 1986.

[PHP95] Pados, D.A., K. W. Halford, and P. Papantoni–Kazakos, "Distributed Binary Hypothesis Testing with Feedback," *IEEE Trans. on Systems, Man, and Cybernetics*, Vol. SMC-25, No. 1, pp. 21–42, Jan. 1995.

[Poo88] Poor, H.V., *An Introduction to Signal Detection and Estimation*, Springer–Verlag, New York, 1988.

[PoT90] Polychronopoulos, G., and J.N. Tsitsiklis, "Explicit Solutions for Some Simple Decentralized Detection Problems," *IEEE Trans. on Aerospace and Elect. Syst.*, Vol. AES-26, pp. 282–292, Mar. 1990.

[PPK93] Pete, A., K.R. Pattipati, and D.L. Kleinman, "Team Relative Operating Characteristic: A Normative-Descriptive Model of Team Decision making," *IEEE Trans. on Systems, Man, and Cybernetics*, Vol. 23, No. 6, pp. 1626–1648, Nov./Dec. 1993.

[PPK94a] Pete, A., K.R. Pattipati, and D.L. Kleinman, "Optimization of Detection Networks with Multiple Event Structures," *IEEE Trans. on Automatic Control*, Vol. 39, No. 8, pp. 1702–1707, Aug. 1994.

[PPK94b] Pados, D.A., P. Papantoni-Kazakos, D. Kazakos, and A.G. Koyiantis, "On-line Threshold Learning for Neyman–Pearson Distributed Detection," *IEEE Trans. on Systems, Man, and Cybernetics*, Vol. 24, No. 10, pp. 1519–1531, Oct. 1994.

[Rad62] Radner, R., "Team Decision Problems," *Annals of Mathematical Statistics*, 33, pp. 857–881, 1962.

[Rao91] Rao, N.S.V., "Computational Complexity Issues in Synthesis of Simple Distributed Detection Networks," *IEEE Trans. on Systems, Man, and Cybernetics*, Vol. 21, No. 5, pp. 1071–1081, Sept./Oct. 1991.

[Rei87] Reibman, A.R., "Performance and Fault-Tolerance of Distributed Detection Networks," Ph.D. Dissertation, Duke University, 1987.

[ReN87a] Reibman, A.R., and L.W. Nolte, "Optimal Detection and Performance of Distributed Sensor Systems," *IEEE Trans. on Aerospace and Elect. Syst.*, Vol. AES-23, pp. 24–30, Jan. 1987.

[ReN87b] Reibman, A.R., and L.W. Nolte, "Design and Performance Comparison of Distributed Detection Networks," *IEEE Trans. on Aerospace and Elect. Syst.*, Vol. AES-23, pp. 789–797, Nov. 1987.

[ReN90] Reibman, A.R., and L.W. Nolte, "Optimal Design and Performance of Distributed Signal Detection Systems with Faults," *IEEE Trans. on Acoustics, Speech, Signal Proc.*, Vol. 38, No. 10, pp. 1771–1783, Oct. 1990.

[Roh83] Rohling, H., "Radar CFAR Thresholding in Clutter and Multiple Target Situations," *IEEE Trans. on Aerospace and Elect. Syst.*, Vol. 19, pp. 608–621, July 1983.

[RWB96] Rago, C., P. Willett, and Y. Bar–Shalom, "Censoring Sensors: A Low–Communication–Rate Scheme for Distributed Detection," *IEEE Trans. on Aerospace and Elect. Syst.*, Vol. 32, No. 2, pp. 554–568, April 1996.

[Sad86]　　Sadjadi, F.A., "Hypothesis Testing in a Distributed Environment," *IEEE Trans. on Aerospace and Elect. Syst.,* Vol. AES-22, pp. 134–137, March 1986.

[SaR83]　　Sarma, V.V.S., and K.A. Gopala Rao, "Decentralized Detection and Estimation in Distributed Sensor Systems," *Proc. 1983 IEEE Systems, Man and Cybernetics Conf.,* pp. 438–441, Vol. 1, 1983.

[ShP92]　　Shalaby, H.M.H., and A. Papamarcou, "Multiterminal Detection with Zero-Rate Data Compression," *IEEE Trans. on Info. Theory*, Vol. 38, No. 2, pp. 254–267, March 1992.

[ShP93]　　Shalaby, H.M.H., and A. Papamarcou, "A Note on the Asymptotics of Distributed Detection with Feedback," *IEEE Trans. on Info. Theory*, Vol. 39, No.2, pp. 633–640, March 1993.

[ShP94]　　Shalaby, H.M.H., and A. Papamarcou, "Error Exponents for Distributed Detection of Markov Sources," *IEEE Trans. on Info. Theory*, Vol. 40, No. 2, pp. 397–408, March 1994.

[Sri86a]　　Srinivasan, R., "Distributed Radar Detection Theory," *IEE Proc., Part F*, 133(1), pp. 55–60, January 1986.

[Sri86b]　　Srinivasan, R., P. Sharma, and V. Malik, "Distributed Detection of Swerling Targets," *IEE Proc., Part F*, 133(7), pp. 624–629, 1986.

[Sri86c]　　Srinivasan, R., "A Theory of Distributed Detection," *Signal Processing,* 11, pp. 319–327, 1986.

[Sri90a]　　Srinivasan, R., "The Detection of Weak Signals Using Distributed Sensors," *Proc. of the Conf. on Info. Sciences and Systems*, Princeton, March 1990.

[Sri90b]　　Srinivasan, R., "Distributed Detection with Decision Feedback," *IEE Proc.*, Vol. 137, Pt F, No. 6, pp. 427-432, Dec. 1990.

[SRV95]   Srinath, M.D., P.K. Rajasekaran, and R. Viswanathan, *An Introduction to Statistical Signal Processing with Applications*, John Wiley & Sons, New York, 1995.

[Swa93]   Swaszek, P.F., "On the Performance of Serial Networks in Distributed Detection," *IEEE Trans. on Aerospace and Elect. Syst.*, Vol. 29, No. 1, pp. 254–260, Jan. 1993.

[SwW95]   Swaszak, P.F., and P. Willett, "Parley as an Approach to Distributed Detection," *IEEE Trans. on Aerospace and Elect. Syst.*, Vol. 31, No. 1, pp. 447–457, Jan. 1995.

[Tan90]   Tang, Z.-B., "Optimization of Detection Networks," Ph.D. Dissertation, University of Connecticut, 1990.

[TeH87]   Teneketzis, D., and Y.-C. Ho, "The Decentralized Wald Problem," *Info. and Control*, Vol. 73, pp. 23–44, 1987.

[Ten82]   Teneketzis, D., "The Decentralized Quickest Detection Problem," *Proc. 21st IEEE Conf. on Decision and Control*, pp. 673–679, Fort Lauderdale, FL., 1982.

[TeV84]   Teneketzis, D., and P. Varaiya, "The Decentralized Quickest Detection Problem," *IEEE Trans. on Automatic Control*, Vol. AC-29, pp. 641–644, July 1984.

[TeS81a] Tenney, R.R., and N.R. Sandell, Jr., "Detection with Distributed Sensors," *IEEE Trans. on Aerospace and Elect. Syst.*, Vol. AES-17, pp. 98–101, July 1981.

[TeS81b] Tenney, R.R., and N.R. Sandell, Jr., "Strategies for Distributed Decisionmaking," *IEEE Trans. on Systems, Man, and Cybernetics*, Vol. SMC-11, pp. 527–537, August 1981.

[ThO92]   Thomopoulos, S.C.A., and N.N. Okello, "Distributed Detection with Consulting Sensors and Communication Cost," *IEEE Trans. on Automatic Control*, Vol. 37, No. 9, pp. 1398–1405, Sept. 1992.

[TPK91a] Tang, Z.-B., K.R. Pattipati, and D.L. Kleinman, "An Algorithm for Determining the Decision Thresholds in a

Distributed Detection Problem," *IEEE Trans. on Systems, Man, and Cybernetics*, Vol. SMC-21, pp. 231–237, Jan./Feb. 1991.

[TPK91b] Tang, Z.-B., K.R. Pattipati, and D.L. Kleinman, "Optimization of Detection Networks: Part I — Tandem Structures," *IEEE Trans. on Systems, Man, and Cybernetics*, Vol. SMC-21, No. 5, pp. 1044–1059, Sept./Oct. 1991.

[TPK92] Tang, Z.-B., K.R. Pattipati, and D.L. Kleinman, "A Distributed M-ary Hypothesis Testing Problem with Correlated Observations," *IEEE Trans. on Automatic Control*, Vol. 37, No. 7, pp. 1042–1046, July 1992.

[TPK93] Tang, Z.-B., K.R. Pattipati, and D.L. Kleinman, "Optimization of Detection Networks: Part II — Tree Structures," *IEEE Trans. on Systems, Man and Cybernetics*, Vol. 23, No. 1, pp. 211–221, Jan./Feb. 1993.

[TsA84] Tsitsiklis, J.N., and M. Athans, "Convergence and Asymptotic Agreement in Distributed Decision Problems," *IEEE Trans. on Automatic Control*, Vol. 29, No. 1, pp. 42–50, Jan. 1984.

[TsA85] Tsitsiklis, J.N., and M. Athans, "On the Complexity of Decentralized Decision Making and Detection Problems," *IEEE Trans. on Automatic Control*, Vol. AC-30, pp. 440–446, May 1985.

[Tsi86] Tsitsiklis, J.N., "On Threshold Rules in Decentralized Detection," *Proc. 25th IEEE Conf. on Decision and Control*, pp. 232–236, Vol. 1, Athens, Greece, 1986.

[Tsi88] Tsitsiklis, J.N., "Decentralized Detection by a Large Number of Sensors," *Mathematics of Control, Signals, and Systems*, Vol. 1, pp. 167–182, 1988.

[Tsi93a] Tsitsiklis, J.N., "Decentralized Detection," in *Advances in Statistical Signal Processing*, Vol. 2, pp. 297–344, 1993.

[Tsi93b]   Tsitsiklis, J.N., "Extremal Properties of Likelihood-Ratio Quantizers," *IEEE Trans. on Commun.*, Vol. 41, No. 4, pp. 550–558, April 1993.

[TVB87a]   Thomopoulos, S.C.A., R. Viswanathan, and D.K. Bougoulias, "Optimal Decision Fusion in Multiple Sensor Systems," *IEEE Trans. on Aerospace and Elect. Syst.*, Vol. AES-23, pp. 644–653, Sept. 1987.

[TVB87b]   Thomopoulos, S.C.A., R. Viswanathan, and D.C. Bougoulias, "Globally Optimum Computable Distributed Decision Fusion," *Proc. 26th IEEE Conference on Decision and Control*, pp. 1846–1847, 1987.

[TVB89]   Thomopoulos, S.C.A., R. Viswanathan, and D.K. Bougoulias, "Optimal Distributed Decision Fusion," *IEEE Trans. on Aerospace and Elect. Syst.*, Vol. AES-25, pp. 761–765, September 1989.

[Une93]   Üner, M.K., "Conventional and Distributed CFAR Detection in Nonhomogeneous Background," Ph.D. Dissertation, Syracuse University, Dec. 1993.

[UnV96]   Üner, M.K., and P.K. Varshney, "Distributed CFAR Detection in Homogeneous and Nonhomogeneous Backgrounds," *IEEE Trans. on Aerospace and Elect. Syst.*, Vol. 31, pp. 84–97, Jan. 1996.

[Van68]   Van Trees, H.L., *Detection , Estimation, and Modulation Theory*, Vol. 1, John Wiley & Sons, New York, 1968.

[VAT88]   Viswanathan, R., A. Ansari, and S.C.A. Thomopoulos, "Optimal Partitioning of Observations in Distributed Detection," Presented at IEEE Int. Symp. Info. Theory, Kobe, Japan, 1988.

[VBP93]   Veeravalli, V.V., T. Basar, and H.V. Poor, "Decentralized Sequential Detection with a Fusion Center Performing the Sequential Test," *IEEE Trans. on Info. Theory*, Vol. 39, No. 2, pp. 433–442, March 1993.

[VBP94a] Veeravalli, V.V., T. Basar, and H.V. Poor, "Minimax Robust Decentralized Detection," *IEEE Trans. on Info. Theory*, Vol. 40, No. 1, pp. 35–40, Jan. 1994.

[VBP94b] Veeravalli, V.V., T. Basar, and H.V. Poor, "Decentralized Sequential Detection with Sensors Performing the Sequential Test," *Math, Control, Signals & Systems*, Vol. 7, pp. 292–305, 1994.

[Vee92a] Veeravalli, V., "Topics in Decentralized Detection," Ph.D. Dissertation, University of Illinois, April 1992.

[Vee92b] Veeravalli, V.V., "Comments on Decentralized Sequential Detection," *IEEE Trans. on Info. Theory*, Vol. 38, No. 4, pp. 1428–1429, July 1992.

[ViA89a] Viswanathan, R., and A. Ansari, "Distributed Detection of a Signal in Generalized Gaussian Noise," *IEEE Trans. on Acoustics, Speech, and Signal Proc.*, Vol. 37, No. 5, pp. 775–779, May 1989.

[ViA89b] Viswanathan, R., and V. Aalo, "On Counting Rules in Distributed Detection," *IEEE Trans. on Acoustics, Speech, and Signal Proc.*, Vol. 37, No. 5, pp. 772–775, May 1989.

[VTT88] Viswanathan, R., S.C.A., Thomopoulos, and R. Tumuluri, "Optimal Serial Distributed Decision Fusion," *IEEE Trans. on Aerospace and Elect. Syst.*, Vol. AES-24, pp. 366–375, July 1988.

[Wal47] Wald, A., *Sequential Analysis*, John Wiley & Sons, New York, 1947.

[WAV94] Willett, P., M. Alford, and V. Vannicola, "The Case for Like-Sensor Predetection Fusion," *IEEE Trans. on Aerospace and Elect. Syst.*, Vol. 30, No. 4, pp. 986–1000, Oct. 1994.

[WaW89] Warren, D., and P. Willett, "Optimal Decentralized Detection for Conditionally Independent Sensors," *Proc. American Control Conf.*, pp. 1326–1329, 1989.

[WiW88] Willett, P.K., and D.J. Warren, "Decentralized Detection: An Information Theoretic Approach," *Proc. 22nd Conf. on Info. Sciences and Systems*, Princeton, March 1988.

[WiW91] Willett, P., and D. Warren, "Decentralized Detection: When are Identical Sensors Identical," *Proc. Conf. on Info. Sciences and Systems*, 1991, pp. 287–292.

[WiW92] Willett, P., and D. Warren, "The Suboptimality of Randomized Tests in Distributed and Quantized Detection Systems," *IEEE Trans. on Info. Theory*, Vol. 38, No. 2, pp. 355–361, March 1992.

[WWR89] Warren, D., P. Willett, and R. Rampertab, "Shannon's Information in Decentralized Signal Detection," *Proc. 23rd Conf. on Info. Sciences and Systems*, Baltimore, March 1989.

Index

Adaptive threshold techniques, 25
Ali–Silvey distances, 110
AND rule, 59, 197-198
Asymptotic probability of error, 134
Asymptotic results, 93
Asymptotically optimum solution, 93
Asynchronous decisions, fusion rule for, 68-70

Backward induction, 218, 229
Bahadur–Lazarfeld expansion, 71
Bayes risk function, 8, 38, 51, 74, 121
Bayesian detection theory, 7-14
Bayesian hypothesis testing for parallel fusion network, 74
Bernoulli random variable, 105
Binary hypothesis testing problem, 6-7, 37, 166, 217

Cell averaging CFAR processor, 28-30, 196-199
CFAR (constant false alarm rate) detection, 24-32
    distributed, 191-206
CFAR detectors, 194
CFAR processor, 25
    cell averaging, 28-30, 196-199
    order statistics, 30-32, 199-206
Clutter-edge case, 202
Communication matrix, 159
    definition generalized, 159
    for parallel fusion network, 138-139
    for serial network, 137-138
Computational algorithms, 101-105
Concavity of team ROC, 87-89
Conditionally independent local observations, 80-89
Constant false alarm rate, *see* CFAR *entries*
Convexification of Lagrangian, 183
Correlated decisions, fusion rule for, 71-72
Cost functions
    entropy-based, 234
    logarithmic, 241-242

Data transmission protocols, 152-159
Decentralized Neyman–Pearson tests, 180
Decision-making problems, 1-4
Decisions
    asynchronous, fusion rule for, 68-70
    correlated, fusion rule for, 71-72

274   Index

Dependent observations, 110-111
Dependent randomization, 89, 180, 182
Detection
    distributed, *see* Distributed detection
    entropy-based, 234
    global and local probability of, 190
    locally optimum, 32-35
    minimax, 14-16
    minimum equivocation, 236-238
    multiterminal, with data compression, 245-249
    nonparametric, 114-117
    with parallel fusion network, 72-117
    probability of, 10, 61, 126
    radar signal, 24
    robust, 111-114
    sequential, 18-24
    weak signal, 32-35
Detection networks, 36
    with feedback, 139-159
    generalized formulation for, 159-178
    non-tree, 161
Detection systems, distributed, 1-4
Detection theory
    correspondence between information theory and, 234-236
    elements of, 6-35
Detectors
    CFAR, 194
    local, *see* Local detectors
    sequencing of, 134-135
    square-law, 25
    time index of, 159, 161
Direct observations, fusion rule with, 67-68
Distributed detection
    based on information theoretic criterion, 234-245
    CFAR, 191-206
    minimum equivocation, 238-245
    Neyman–Pearson, 179-191
    nonparametric, 114-117
    sequential, 216-232
    systems, 1-4
    of weak signals, 206-215
    without fusion, 37-58

Dynamic programming, 219, 229-230

Entropy, relative, 246
Entropy-based cost functions, 234
Entropy-based detection, 234
Equivocation, 234
Error
    asymptotic probability of, 134
    minimum probability of, decision rule, 43
Error exponent, 93
Error exponent function, 246

f-divergences, 110
False alarm, 10
    global and local probability of, 190
    probability of, 10, 61, 126
Feedback
    detection networks with, 139-159
    parallel fusion network topology with, 140-159
    parallel fusion network topology without, 151-152
Finite-horizon problem, 218, 221-222, 228
Fusion center, 3
    parallel topology without, 238-245
    sequential test performed at, 226-232
Fusion rules
    AND, *see* AND rule
    for asynchronous decisions, 68-70
    for correlated decisions, 71-72
    design of, 59-72
    with direct observations, 67-68
    monotonic, 63-65
    optimum, 62-63
    OR, *see* OR rule
    for soft decision case, 65-67

Gamma pdf, 29
Gauss–Seidel algorithm, 111
Gauss–Seidel cyclic coordinate descent algorithm, 101-102
Generalized Neyman–Pearson lemma, 206
Gradient based iterative algorithms, 104

Index   275

Homogeneous background, 26
Hypothesis testing problem binary, 6-7, 37, 166, 217
   components of, 7
   multiterminal, 249

Identical local detectors, 89-101
Independent randomization, 89
Infinite-horizon problem, 218, 221, 222
Information, mutual, 236
Information loss, 234
Information-optimal threshold, 236-238
Information theoretic criterion, distributed detection based on, 234-245
Information theory, 233-249
   correspondence between detection theory and, 234-236
Informational divergence, Kullback–Leibler, 246

J-divergence, 109

K-out-of-N fusion rule, 93-94, 95
Kolmogorov variational distance, 13
Kullback–Leibler informational divergence, 246

Lagrange multiplier method, 33
Lagrange multipliers, 17, 183-184, 195, 196, 208
Lagrangian, 183
   convexification of, 183
Least favorable distributions, 113, 114
Least favorable prior, 14
Level of test, 16
Likelihood ratio partitioning, 182
Local detectors, 177-178
   identical, 89-101
Local observations, conditionally independent, 80-89
Locally optimum detection, 32-35, 206
Logarithmic cost functions, 241-242

M-ary hypothesis testing, 50
MAJORITY rule, 96

Minimax criterion, 15
Minimax detection, 14-16
Minimax robustness, 112, 113
Minimax test, 15
Minimum equivocation detection, 236-238
Minimum probability
   of error decision rule, 43
   of error receivers, 10
Miss, 10
   probability of, 10, 61
Monotonic fusion rules, 63-65
Most powerful test, 16, 32
Multiple target case, 202
Multiterminal detection with data compression, 245-249
Multiterminal hypothesis testing problem, 249
Mutual information, 236

N-out-of-N fusion rule, 95
Neyman–Pearson criterion, 16
Neyman–Pearson detection, distributed, 179-191
Neyman–Pearson lemma, 16, 181
   generalized, 206
Neyman–Pearson tests, 16-18
   decentralized, 180
Non-tree detection networks, 161
Nonhomogeneous background, 32, 202
Nonlinear programming problem, 103
Nonparametric detection, 114-117
NP-completeness, 111

Observations
   conditionally independent local, 80-89
   dependent, 110-111
   direct, fusion rules with, 67-68
   uncertain, 111-114
Optimum fusion rules, 62-63
OR rule, 61, 198-199
Order statistics CFAR processor, 30-32, 199-206
Ordering of detectors, 134-135

Parallel detection network without fusion, 3, 37

Parallel fusion network, 3
  Bayesian hypothesis testing for, 74
  communication matrix for, 138-139
  detection with, 72-117
  serial network versus, 128, 132
  with soft decisions, 107-110
Parallel fusion network topology, 72
  with feedback, 140-159
  without feedback, 151-152
  Neyman–Pearson formulation, 180
  solution for weak signals, 206-215
Parallel topology
  fusion, see Parallel fusion network entries
  without fusion center, 238-245
Partitioning, likelihood ratio, 182
PBPO (person-by-person optimization), 75-80
Person-by-person optimal solution, 103
Person-by-person optimization, see PBPO entries
Poisson random variable, 69

Radar signal detection, 24
Randomization, 17, 87-89
  dependent, 89, 180, 182
  independent, 89
Rate constraint, 247
Rayleigh conditional densities, 47
Rayleigh fading environment, slow, 190
Receiver operating characteristic (ROC), 14, 83
  concave, 182
  team, see Team ROC
Relative entropy, 246
Robust decision rule, 113
Robust detection, 111-114
Robustness, minimax, 112, 113
ROC, see Receiver operating characteristic

Scheduled tests, 180
Sequencing of detectors, 134-135
Sequential detection, 18-24
Sequential probability ratio test (SPRT), 18-24

Sequential test, 217
  performed at fusion center, 226-232
  performed at sensors, 217-226
Serial network, 120-136
  communication matrix for, 137-138
  design of decision rules for, 120-136
  parallel network versus, 128, 132
  two-detector, 127, 135
Serial topology, 3-4
Sign detector, 114-115, 116, 117
Slow Rayleigh fading environment, 190
Slowly fluctuating target, 25
Soft decision case, fusion rules for, 65-67
Soft decisions, parallel fusion network with, 107-110
Spatial independence, 146
SPRT (sequential probability ratio test), 18-24
Square-law detector, 25
Stein's lemma, 246
Stopping time, 217
  optimal, 220
Swerling slowly fluctuating target, 25

Tandem detection network, 120-136
Tandem topology, 3-4
Team ROC
  concave, 182
  concavity of, 87-89
Threshold, information-optimal, 236-238
Threshold strategies, 110
Time index of detectors, 159, 161
Tree configuration, 176
Tree networks, 137-139
Truncated SPRT, 24
Two-detector serial network, 127, 135

Uncertain observations, 111-114

Wald's sequential test, 18-24
Weak signals
  detection of, 32-35
  distributed detection of, 206-215
Wilcoxon detector, 114-115, 116, 117